POPE PIUS XII LIBRARY, ST. JOSEPH COL.
3 2528 07525 1555

D1525831

REINHABITING THE EARTH

REINHABITING THE EARTH

Biblical Perspectives and Eco-Spiritual Reflections

MARY LOU VAN ROSSUM

TRIUMPH™ BOOKS
Liguori, Missouri

Published by Triumph™ Books
Liguori, Missouri
An Imprint of Liguori Publications

Library of Congress Cataloging-in-Publication Data

Van Rossum, Mary Lou.
Reinhabiting the earth : biblical perspectives and eco-spiritual reflections / Mary Lou van Rossum.
p. cm.
Includes bibliographical references and index.
ISBN 0-89243-691-3
1. Nature—Religious aspects—Christianity. 2. Human ecology—Religious aspects—Christianity. 3. Creation—Biblical teaching.
I. Title.
BT695.5.V365 1994
261.8'362—dc20 94-7302
CIP

Printed in the United States of America
First Edition

This book is printed on recycled paper.

CONTENTS

Part Two
THE CHRISTIAN SCRIPTURES

FOREWORD

In this final decade of the twentieth century, we live amid spectacular events. Mountains are being torn apart so their minerals can be plundered. Great tropical forests that took millions of years to bring into being are destroyed in a few short years. Biological species and entire ecosystems that had flourished for millennia are crashing all around us. Oceans of petroleum, the result of untold trillions of organisms that lived in the past, are being drawn from their hidden recesses, used, and the residue thrown into the air or drained into the earth. The great rivers of the world are being dammed for water supply and energy and irrigation.

Human genius in its scientific insight and technological skill is altering the relationship of humans with both the physical and the biological structures and functioning of the world about us. These changes, however beneficial in their immediate use, are for the most part both deleterious and irreversible. We wonder at the possibilities of recovery.

As we search for a world of meaning that could guide us in our actions in this period, we in the Western world turn to the sources that have led us through the perilous moments of the past. Foremost among these sources are the revealed traditions presented to us in that amazing collection of writings that we know as the Bible. These writings, in their great moments, emerge from interior experiences that took place in the deepest realms of the human unconscious, experiences so comprehensive in their significance and so transforming in their consequences that they were perceived as coming from the ultimate source of existence itself.

Foremost among these experiences has been a sense of the mysterious origin and unity of the universe. Nothing in the biblical teachings is given more emphasis than the unity of the world, its emergence from a single source, and its continued dependence upon that source for its existence, its every activity, and its final fulfillment. Because the universe emerges from a single sacred source, it participates in this sacred character of its origin.

The mysterious order of the universe, its fruitfulness, its energy — all these are present from the beginning and ever remain as manifestation of the numinous reality whence they came. Every being in the universe is bonded together with every other being in terms of their origin, their intimate dependence, and their destiny. This unity is seen especially in those glorious moments when the entire community of existence joins in a hymn of praise such as we find in Psalm 96, where the entire order of creation joins in a single voice. "Let the heavens be glad and the earth rejoice / let the sea and what fills it resound; let the plains be joyful and all that is in them / then shall all the trees of the forest exult before the Lord."

In this psalm, as in the great creation psalm 104 and in so many of the other songs of the Bible, the entire order of creation is experienced as a vast celebration, as a grand cosmic ritual, as sheer delight in existence and in praise of that mysterious and ineffable source whence came the stars in the heavens and all the living creatures of Earth. The very purpose of the human is to enter into this celebration with an understanding of its order, appreciation of its beauty, enjoyment of its delights, and obedience to its guidance.

From earliest times, humans have understood this as they arrange their own ritual participation in the seasonal sequence of the natural world, especially in the springtime renewal ceremonies after the decline of winter. Initiation rituals are associated with their stories of how the universe came into being, and how humans enter into this larger order of the universe. Human activities are consistently inserted within the spatial context of the universe by constant references to the four directions and to the above and below of things. This is necessary since the unity of the human with the entire universe must always be observed.

Every activity is a participated activity. Indeed, the universe itself is the larger self of every being. To bring the smaller self of the individual into relationship with the larger human community and then finally with the entire universe community is to achieve personal integrity and final fulfillment. It is to establish the identity between the smaller self and the great self.

This sense of participated activity throughout the universe we find in the Sioux Indian Black Elk when he had his great vision at the age of nine. In that vision, he was taken into the sky and saw all the earth below. When the great drama he witnessed in heaven was over, he observed that all the earth was caught up in the celebration. Every being of Earth joined in the great ritual dance, the birds and the animals, the leaves on the trees, the waters in the streams, the flowers in the fields, the men and the maidens — all danced to the song of the great stallion that appeared in the heavens.

Such intimacy with the natural world is less available to us now as we live in cities amid all the rush of human affairs in a commercial-industrial society. We have learned to shield ourselves from the chill of winter and the heat of summer. We travel across the planet with a certain ease, have foods from all six continents on our tables. We heal diseases that formerly were severe afflictions. We communicate instantly with any point on the planet. Yet, with all this power over the physical dimensions of the planet and over many aspects of our lives, we have in a sense lost the universe.

We have become autistic in relation to the natural world in terms of any meaningful relation or any intimate presence to natural phenomena. As with children who have suffered some psychic trauma and cannot communicate with the outside world or let the outside world into their own psychic space, so we no longer have any extensive communion with the outer world. Because we no longer have this intimacy with the sun and wind and rain, with the fields and forests, with the clouds in the day or the stars at night, we have grown alienated from some of the most profound experiences that humans have been capable of in former times.

These experiences are not incidental or addenda to our human lives. Our loss of the universe is a loss of soul. These experiences are essential to our integral development as humans. Scientific inquiry has confirmed this need that humans have for presence to the natural world. E. O. Wilson, one of our foremost biologists, has indicated that the living world about us is the "matrix in which the human mind originated and is permanently rooted." And our highest faculties evolved "in a biocentric world, not a machine-regulated world." Not only E. O. Wilson but a multitude of other writers are now leading us back into these experiences. This we find in the world of Morris Berman, *The Reenchantment of the World*. Also, there is the work of Robert Keck entitled *Sacred Eyes*. Books of prayers and liturgies have been published to assist in our reintegration of the human community into the larger sacred community. Poets such as Mary Oliver are awakening us to these primordial experiences. Above all, perhaps, the long list of naturalists in the last half of the twentieth century are restoring our sensitivity to the larger-life community of the planet.

We began to reinhabit the earth. We awaken to the earth in a mutually enhancing relationship. This is something more than a type of calculated adjustment. It is the new sense of intimacy indicated in that superb expression of E. O. Wilson, *Biophilia,* a sense of the mutual presence that naturally exists between the human and the entire bioregion in which we live. The question is not simply that we depend for our physical lives on the food and air and water of our bioregion. Our souls are awakened by the impressions that we receive from without, from the stars in the heavens and the mountains and valleys and rivers and all the living forms.

Divine reality itself comes to us through our experience of the vastness of the oceans and the feel of the wind, through the song of the mockingbird and the howl of the wolf, through the hurricanes that sweep up from the south Atlantic and earthquakes that shake the Pacific Coast and the floods that pour down through the great central valley of the North American continent.

Gratefully, we are beginning to recover this loss of our interior world. We begin once again to integrate our lives with the great liturgy of the seasonal cycle, with the diurnal liturgy of the day and night. We awaken to a realization that any authentic religion or any authentic spirituality must be an expression of the entire sacred community, which includes the entire bioregional community. For ourselves, this in its larger dimensions is the North American continent. For this is the locus wherein we ourselves and the divine and the natural world experience the most profound fulfillment.

With Mary Lou Van Rossum, we find that this insight is the deepest teaching of the biblical literature. There we find this intimate relationship of the three orders of the universe, in the beginning with Genesis and its fulfillment in the last pages of the Book of Revelation of Saint John. This is the teaching of Isaiah the prophet of the Messianic times when the wolf and the lamb, the leopard and the kid, the calf and the lion, the cow and the bear, all dwell together in the presence of the little child who plays by the cobra's den.

A great reconciliation is taking place in our times, the reconciliation of the fourfold wisdom that can lead us into the magnificence of the Ecozoic era. This fourfold wisdom is the wisdom of the traditions, the wisdom of women, the wisdom of indigenous peoples, and the wisdom of science. None of these can any longer do without the others. *Reinhabiting the Earth* is a superb beginning from the view of the ancient Scriptures of our Western world.

THOMAS BERRY

PREFACE

We find ourselves today at the edge of a great transformation of consciousness. Wherever we look, we see evidence of breakdown and breakthrough, of chaos and creativity. As we move toward a new millennium and the spontaneous emergence of a new order of being, we find ourselves hovering tenuously between fear and trust, despair and hope.

This crisis of emergence is not a new reality for the human community. It is something that is inherent in the order of creation and that has been going on, in greater or lesser degrees, in the human community since our emergence as humans. What is new in our day is the magnitude of the situation and our responsibility in the situation.

In times past, our ancestors encountered these moments of transformation. Some of them made it through. Others did not. In the emergence of humans, 2–3 million years ago, two of our ancestral hominids did not survive. They were strong and powerful, but they perished. One, however, evolved into *Homo habilis* and learned to speak, to conserve fire, and to fashion primitive tools from bone and wood. In this indeterminate moment of grace, early humans came into being.[1]

Our human ancestors and their companion animals also experienced a tremendous time of crisis and transformation one million years ago in the coming of the Great Ice Age. In this time of intense cooling, the region of the African Sahara dried up, initiating a mass migration of *Homo erectus* and large grazing animals to the north, south, and east. Those who traveled eastward moved out of Africa and settled in Europe and Asia, learning to adapt to new environments and new food sources.

A more recent crisis and transformation occurred 12,000–15,000 years ago as the last period of glaciation came to an end. At this time, global warming and excessive hunting by the ice-age hunters resulted in the extinction of most ice-age mammals. This loss initiated a food crisis that substantially reduced hunting and gathering and intensified fishing, herding, and agriculture. It marked the end of the human community as a hunter-gatherer society and the beginning of the agricultural revolution.

Our Judeo-Christian tradition records another moment of profound change that clearly involves both the physical-material and psycho-spiritual dimensions of a people moving through a process of crisis and transformation. After a dangerous flight from Egypt and a long and perilous journey through the desert, the tribes of Moses arrive at the plains of Moab on the eastern side of the Jordan River. Within the sight of a promised land, flowing with milk and honey, Moses recalls his people to a covenanted life of fidelity and love:

> ... this law which I am laying down for you today is neither obscure for you nor beyond your reach. It is not in heaven, so that you need to wonder, "Who will go up to heaven and bring it down for us, so that we can hear and practice it?" Nor is it beyond the seas, so that you need to wonder, "Who will cross the seas for us and bring it back to us, so that we can hear and practice it?" No, the word is very near to you, it is in your mouth and in your heart for you to put into practice.
>
> Look, today I am offering you life and prosperity, death and disaster.... Choose life, then, so that you and your descendents may live ... (Dt 30:11–19 [NJB])

This call to choose life, which is at the heart of every moment of existence and intensifies in moments of transformation, is the great decision before us now. In our day, it involves not only our own lives and the lives of our descendants, but the survival of all life-forms on the planet. Today, we hold in our hands and in our hearts the choice of life or death, prosperity or disaster, for every living species and life-system on the earth.

Today, choosing life requires an exodus from our present social order based on anthropocentrism, militarism, consumerism, and technomania to an ecologically conscious global community based on biocentric and geocentric perspectives. It requires a compassionate and disciplined lifestyle that lives within the resource boundaries of the natural world and honors the unity and diversity within the community of creation. Today, choosing life requires a perceptive sensitivity to the ongoing process of creation and a passionate involvement with the natural world. It requires a fidelity to the Mysterious Love at the heart of creation, and a lived congruence with our highest intimations of truth, beauty, justice, compassion, and community.

This present crisis of the earth community and the need for an eco-spiritual consciousness is the compelling ground of my writing. Though I realize that these issues of spirituality and ecology may be addressed from many perspectives and through many spiritualities and religious traditions, I address them through the Judeo-Christian tradition. I do this not because I believe that it holds all the answers to our present situation but because it is my own tradition and the only one in which I hold sufficient grounding for the task. It is my intent that through this writing I may enter into and contribute to the spiritual dialogue of the wider community of creation.

In doing this work, I have explored the Judeo-Christian Scriptures and reflected on ways in which they speak an ecological wisdom to our age. In my writing, I have included biblical awarenesses and ecological reflections in each chapter, moving through the Scriptures in the seven major areas of biblical writing — the Prehistory of Genesis, the Law, the Prophets, the Wisdom Literature, the Gospels, the New Testament Letters, and the Book of Revelation. I have chosen to ground my work in these seven major areas not only to reflect the organic structure of the Scriptures, but also to reflect the seven days of creation in Genesis, the continuing significance of the number seven throughout the Scriptures, and its powerful significance in the Book of Revelation.

I have begun each chapter with an exegetical commentary on the contemporary understanding of that body of writing as a whole and concluded each chapter with ecological reflections that illuminate biblical themes and address contemporary concerns.

The first of these contemporary concerns is ecological sensitivity. It expresses our need to live in harmony with the earth and the earth community through the pursuit of a consciousness and a lifestyle that honor the sacred dimensions of the earth and the communities of the earth. The second concern is an evolutionary perspective. It expresses the awareness that creation continues, that we live in an evolving universe where all things evolve through time, even humans and human consciousness. The third concern is that of cultural transformation. It expresses the reality that creation moves in evolutionary bounds and that there are disjunctive times of mutation or qualitative transformation that occur not only in the natural world, but also in the inner realms of our psycho-spiritual lives and the outer realms of our cultural milieus. These three concerns emerge randomly in my reflections in response to the biblical material. There is a natural movement within the Scriptures, however, from ecological sensitivity to cultural transformation, and I believe that my reflections follow that movement.

As many readers may not be familiar with the contexts of the biblical writings, I have introduced this book with a brief and rather intense sweep of biblical history. As much of this history is based on the work of Robert B. and Mary P. Coote, I would recommend their book *Power, Politics and the Making of the Bible* to anyone who wishes to pursue this history in more depth. The Introduction also recounts the development of the three major bodies of writing in the Hebrew Scriptures — the Law, the Prophets, and the Wisdom Literature — and the three major bodies of writing in the Christian Scriptures — the Gospels, the New

Testament Letters, and the Book of Revelation. This material serves as an introduction to both the form and the content of the chapters that follow.

In Chapter One, "The Prehistory of Genesis," I look to the stories of Creation and the Flood as the primal myths of our Western society and explore them through contemporary exegetical understandings. In the reflections, I draw out five themes that originate in these primitive stories and flow through the later Scriptures. These themes are the sovereignty of God and the unity of creation, the inherent value of nonhuman creation, human ecology, the Sabbath, and covenant. Everything that follows is related to one or more of these themes.

Chapter Two, "The Law," explores the legal codes of Israel as expressions of the way in which the people of Israel understood their call to live in fidelity and obedience to Yahweh within the relational context of covenant. The reflections look to these laws as patterns that may inform our relations with the living and nonliving systems of earth, with our companion animals, and with the foreigners and strangers in our midst.

In Chapter Three, "The Prophets," I explore the major concerns of the prophetic writings, focusing on fidelity to Yahweh, social justice and concern for the poor, and the concept of a continuing creation in which Yahweh continues to bring forth new realities from the chaos and destruction that follow infidelity and failure. In the reflections, I explore the understanding of Yahweh as the Holy One of Israel, the continuing cry of the poor in both human and nonhuman creation, and the continuing transformation that is inherent in the creative will of God and, subsequently, in creation itself. I also assert my belief that this creative will empowers us to hold on to hope in the midst of darkness and despair and to seek congruence with Ultimate Reality through repentance and return.

Chapter Four, "The Wisdom Literature," explores the origins and the perceptions of wisdom within the traditions of Israel. It reveals that the wisdom tradition does not draw upon the law or the prophets, upon covenant or cult, but looks for wisdom in the ordinary experiences of our everyday lives. The wisdom reflections center on the mystery of God and the creativity that is inherent in the world, on the feminine face of God that is found in the Wisdom Woman of these writings, and on how we might live wisely on the earth.

In Chapter Five, "The Gospels," I follow the emergence of these writings within the early church, and the varying perspectives of the four evangelists — Matthew, Mark, Luke, and John. In these reflections, I look to Jesus as the initiator of the New Creation and the bearer of a transformed consciousness, as the one who fulfills the law, the prophets, and the wisdom literature of the Hebrew Scriptures. These reflections also explore justice, compassion, and inclusion in the community of creation through the prophetic exhortation to "act justly, love tenderly, and walk humbly with one's God" (Micah 6:4), and on the commands to forgive and to love one's enemy through nonviolence and living harmlessly on the earth.

Chapter Six, "The New Testament Letters," follows the developing theologies of the early church through the writings of Saint Paul, the disciples of Paul, and other teachers in the Christian communities of the first century. These theologies center on the person and work of Jesus, on living in mutual love and service within the Christian communities, and on living peaceably in the midst of pluralistic, and sometimes hostile, societies. The reflections look to the hymns of the Cosmic Christ not only as reflections on the sovereignty of God and the unity of creation, but also as expressions of the human voice in the larger hymn of the universe. These reflections also invite us to carry on the mutual love and service of the Christian community in the pluralistic communities of our human cultures, the mixed communities of our own bioregions, and the known and unknown realms of the universe.

In Chapter Seven, "The Book of Revelation," I explore the apocalyptic genre, looking to its exilic roots in the visions of Ezekiel and following its movement into the final book of the Christian Scriptures. In the reflections, I explore images of endtime through biblical and post-biblical writings, including both dualistic disaster visions and ultimate reconciliation visions. This exploration also invites us to come to terms with the existence of our personal, collective, and cosmic shadows. I conclude this final chapter with a reflection on cultural transformation and visions and dreams that may help us move through our present time of crisis into a time of profound planetary transformation.

While the biblical studies in my writing are based on contemporary exegetical studies, the reflections are, quite literally, "reflections" and are wholly heuristic. They arise from my own meditative interaction with the biblical texts and reflect my own intuitive understandings, perceptive experiences, and academic studies. It is my hope that in their heuristic origins they will find resonances in the heart of the reader.

ACKNOWLEDGMENTS

There have been two enduring loves in my life. The first of these is my love for the natural world. The second is my love for the Judeo-Christian Scriptures.

I have always been enchanted by the natural world, the grandeur of the heavens and the loveliness of the earth. I have found joy in the high-flying clouds, and I have been brought to silence by the night sky. The mountains and hills, the rivers and trees, have sustained and nurtured my life. Wildflowers and lilacs, mourning doves and squirrels, have companioned me along my way. The whole universe has been a place of wonder and awe for me.

The Judeo-Christian Scriptures have also befriended me, in my solitude and in my experiences of community. They have comforted me in my afflictions and set me free when I have been imprisoned by my cultural milieu or the turmoil of my own thoughts. They have given sight in my experiences of blindness and opened my ears when I have been deaf to the voices of the worlds within me and around me. Throughout my life, I have always found God in both the natural world and in the Scriptures.

To bring these loves together in an extended reflection is to explore something that has long been part of my ordinary life. It has been an enormous challenge and a very great joy. Because I have been immersed in the study of both the Scriptures and contemporary ecological issues for many years, it is difficult at times to remember where I have come upon some of the knowledge

I have acquired and to give the appropriate credits or citations for what I am writing. Much of what I have written has come through my own research and reflection as a scholar, homilist, and writer, but most of what I have written is ultimately based on the scholarship of other authors. Often it is an amalgam of several authors whom I can no longer distinguish or trace, or material that may be unknown to the casual reader but is understood as common knowledge within the academic community of biblical scholarship or deep ecology. It is with these caveats that I make my acknowledgments, and to these unnamed and unnumbered men and women who carry and enhance the tradition that I express my first gratitude.

In the area of Scripture, I am especially indebted to my seminary professors, who not only shared their own wisdom and knowledge but also empowered me in the art of biblical scholarship. I am grateful to Richard J. Sklba, who deepened and enhanced my love for the Hebrew Scriptures and whose teachings communicated and evoked great enthusiasm and joy. I am also grateful to Thomas Suriano for the warmth, light, and excitement he brought to his teaching of the Christian Scriptures, and for his uncommon integrity that incarnates the gospel life of peace, justice, and compassion. I am also grateful to Tom for leading me and my sister and brother seminarians on an unforgettable pilgrimage to Israel and for opening up the sights and sounds and feelings of the biblical world for us.

In the area of eco-spirituality, I am grateful, first and foremost, to my father, who shared his love for the natural world with me. A true child of the universe, he introduced me to the wildness of the natural world and taught me everything he knew about the night sky. As a musician and a singer, he also introduced me to the mysterious world of the Scriptures through the solemn liturgies and the Gregorian chant of the Roman Catholic Church.

I am also grateful to my grandfather, whose solitary labor and love for the great garden he created, cultivated, and cared for continually leads me to reflect on the ancient and solitary Gardener who hovers over the great garden planet s/he has brought into being.

In the immediacy of this work, I am indebted to Thomas Berry and Brian Swimme, who have led me to a new and deeper understanding of the beauty and terror of the universe and the fragility of the earth and the earth community. Their lectures and writings inform much of my writing in this present work. I am also indebted to Nancy Owens, Kevin Sharpe, Thomas Berry, James Conlon, Jean Troy Smith, and Armand Alcazar, who have contributed their wisdom and insight and have always been there to listen, care, and support me in this work.

I am also thankful for the staff and resources of the Flora Lamson Hewlett Library at the Graduate Theological Union in Berkeley, California. Their enormous open collection of theological works has enabled me to explore the thoughts of a multitude of authors, and their many reference shelves have been a rich and continual source of information through biblical dictionaries, encyclopedias, and commentaries.

Last but not least, I am grateful to Carol and Roger, the keepers of the land, over whose garage and in whose wildlife refuge I have lived and written this book. I am grateful for their uncommon love for wilderness and wild things and for their allowing me to share life with them and the creatures of their land. Though they would not understand my gratitude, I am grateful to the cats and dogs and goats, to the fox and deer and raccoon, and to the countless winged beings that have enriched my life — hummingbirds, mockingbirds, jays, hawks, mourning doves, and great horned owls. All these have been my teachers, and I have learned much from them about living simply and compassionately in the community of creation.

Throughout this work, I have used the New American Bible with the Revised New Testament as my primary biblical text. This Bible was copyrighted in 1986 by the Confraternity of Christian Doctrine in Washington, D.C. The illustration from this Bible was copyrighted by Catholic Book Publishing Company in New York in 1987. In addition to this text, I have used the New Jerusalem Bible and the New Revised Standard Version Bible as alternative texts. The New Jerusalem Bible was copyrighted by Doubleday and Company, Inc., in 1985. The New Revised Standard Version Bible was copyrighted by the Division of Christian Education of the National Council of the Churches of Christ in the United States of America in 1989. All biblical quotations have been taken from the New American Bible unless they are noted as "NJB" or "NRSV."

INTRODUCTION

The Judeo-Christian Scriptures emerge out of the dim recesses of sand and time. They come shimmering across the ages from ancient Near Eastern lands and draw us into the deep past of our ancestors-in-faith. If we choose, we can follow them back through time and memory, through words and stones, searching for what we can know and for what is ultimately unknowable. And, we can build from them a coherent history, a stage on which our ancestors come to life again, on which they sing their songs and plow their fields, on which they grow in faith or perish in folly. It is this history, this journey in words and stones, with which I begin.

Early Israel[1]

In the sixteenth century B.C.E., the territory that was to become Israel was a rich and prosperous land. Its urban centers were fortified and populated by a wealthy and powerful elite. Its agricultural lands were well cultivated and occupied by poorer workers who gathered in affinity groups centered on family, village, or tribe. Like other people of the ancient Near East, the people of ancient Palestine engaged in a wide variety of arts and crafts as well as politics and war. They raised wheat and barley, olives and grapes, sheep and goats. Each tribe or affinity group was protected by its sheikh or local strongmen and had its own cult made up of gods, goddesses, spirits, and demons whom it revered, patronized, or appeased.

In all these ways, ancient Palestine differed little from other lands in the ancient Near East. One reality, however, was different. Ancient Palestine lay in the fertile crescent at the crossing of trade routes, both land and sea, that linked the eastern Mediterranean lands of Crete and Greece, the ancient Near Eastern lands of Anatolia and Mesopotamia, and the African nations of Ethiopia and Egypt. Because of this location at the heart of the trade routes, the urban centers were quite cosmopolitan, with a variety of inhabitants and languages, and the land itself was coveted by surrounding nations that fought to control the trade routes.

In these ancient times, Palestine served as a buffer zone between the Hittite Empire to the north and the Egyptian Empire to the south, and often experienced incursions from both of these warring nations as they attempted to bring Palestine into subjection and expand the base of their power and trade.

About 1550 B.C.E., the Egyptians, having overthrown their Hyksos overlords, began the era of their New Kingdom and initiated a series of periodic invasions into Palestine that continued for four hundred years. These invasions intensified in 1468 B.C.E., when the reigning pharaoh, Tuthmose III, began a brutal series of annual invasions that lasted for almost forty years.

In the first year, Tuthmose III moved up along the coastlands, defeating the Asian coalition at Meggido and taking over the hub of the main trade routes across the land. In the years that followed, he continued to move deeper into Asia until he had reached the northern territories of Syria. In each of these invasions, he plundered cities, set up military garrisons and tribute systems, and carried away the inhabitants and riches of the lands. These intense assaults brought the once-prosperous territory into subjection, slavery, and impoverishment.

Over the years, the Egyptian-occupied land in southern and coastal Palestine became known as Canaan and disintegrated into a survivalist land of the displaced and the dispossessed, of renegades and warring tribes. As Egypt was never able to establish uniform control in the land, the Egyptians turned much of its administration over to local strongmen and foreign mercenaries, and some of the captives carried into Egypt were trained and returned to Palestine to oversee lands and exact tributes. In this way, the relationship between Egypt and Palestine became complex and ambiguous, and all forms of power became tenuous and subject to violent reversals.

About 1150 B.C.E. the warring powers of Egypt and the ancient Near East receded from history. The reasons for this are unclear, but historians think a famine may have swept through the land or that the "sea people" of the Mediterranean may have invaded and conquered these lands.[2] As these old powers declined, the Philistines, who had governed the southern coastlands for Egypt, began a struggle for power in Palestine. These new invaders moved across the coastal foothills, into the hill country, and on to the Jordan Valley, taking over the trade routes as they advanced. In the hill country, tribal chieftains struggled against them and eventually came together as a confederation.

Early Israel grew out of the ravages of this Egyptian conquest and the Philistine invasion. It emerged in the thirteenth century B.C.E. as a tribal military force that revered El, the tribal chief of the gods. It grew to a confederation from the local tribes and villages that remained, from the dispossessed and the renegades who had fled into the hills and from ancient Palestinian families who had been repatriated by the Egyptians or who had returned after the decline of the Egyptian Empire. It became a tribal land of village people engaged in agriculture and pastoral nomads who tended their flocks and herds. Its history of struggle formed the core of its character as a nation.

The Hebrew Scriptures

The Scriptures, as we know them, emerged among these tribes of Israel. They began as stories and songs recounting the memories of their ancestors, the lives of their heroes, and their early struggles to inhabit the land. These early stories were probably told both within the family and by professional storytellers and used to entertain, teach, and edify as well as maintain and strengthen tribal community. The earliest proverbs, the Moses stories of the Exodus and Covenant, tales of the ancestors, stories from the book of Judges, the war songs, ancient blessings of Israel, and the Eden and Flood stories were probably among these early oral traditions.[3]

These stories contained history, cultic law, prophetic insights, and folk wisdom. Over the years, they developed into three distinct bodies of writing. The first and most important group of writings for the Israelite people was that of the priests or cult, which contained the law or Torah and the early history of Israel. The second most important, which stood as a critique of the priesthood and the cult, was the writings of the prophetic guilds or the prophetic literature. The third group of writings was the teachings or "writings" that carried the wisdom of the community. As they grew, each literary circle had its own interests and perspectives and its own sphere of influence.

These Scriptures were first committed to a written form by the Yahwist author in the court of David under the direction and most likely the watchful eye of David. This Yahwist tradition continued on and flourished in the reign of Solomon. Its central theme was the Mosaic exodus from the oppression of Egypt and the covenantal relationship with the tribal god El, who was revered under the name of Yahweh. Known for its simple, earthy stories and the use of the name Yahweh, this tradition articulates Israel's self-understanding through its tribal epics. These epics include the garden of Eden and flood stories and the stories of Abraham, Isaac and, perhaps, Jacob, all of which had come into Israel's folk history through cuneiform texts from Mesopotamia.[4] They also include the Mosaic tradition which had come from the pastoral nomads in southern Palestine and the earliest stories of the Judges. To these already existing oral traditions the Yahwist added the psalms of David, the proverbs of Solomon's court, and the stories of David's coming to power and Solomon's reign.

A second tradition, the Elohist tradition, emerged around 900 B.C.E. in the Northern Kingdom of Israel when the United Kingdom fell apart after the death of Solomon. The relationship between the tribes of the more urbane agricultural north and the hilly pastoral south had always been shaky. Their cults and their cultures had differed, and the northern lands, which contained the hub of the trade routes, were more open, fluid, and syncretistic in relating to foreign powers. When Solomon's son Rehoboam ascended to the throne in Jerusalem and intensified the harsh rule of Solomon, the northern leaders rebelled and broke away, establishing their own kingdom with Jeroboam as their king.

Jeroboam and the Northern Kingdom retained the earlier writings of the Southern Kingdom and added material that affirmed and legitimated the division of the kingdom and the reign of Jeroboam. The writings of these years included this later court history, the stories of Elijah and Elisha, and the prophetic works of Amos and Hosea. This Elohist material is distinguished not only by its history and geography of the Northern Kingdom, but by its use of the plural form of the deity in the name Elohim and in its concern with the prophets, the local saints, and their miracles.

When the Northern Kingdom fell to Assyria in 722 B.C.E., these Elohist writings were carried to the south and joined with the writings of the Southern Kingdom. From this composite, a Yahwist-Elohist strand emerged.

The third major strand, known as the Deuteronomic tradition, began in Jerusalem around 700 B.C.E. in the reign of Hezekiah. Wishing to revise the Yahwist-Elohist material in a language and rhetoric more suitable to the time and in support of a centralized government and system of law, Hezekiah assigned his scribe the task of reappropriating the Davidic history and the history of the Northern Kingdom. These writings became the beginnings of the Deuteronomic tradition. They continued on through the fall of Jerusalem and were completed during the time of the Babylonian exile.

The last major strand of the Hebrew Scriptures, the priestly tradition, had its beginning among the Aaronid priesthood during the Babylonian exile when the priestly writer(s) began a revision of the Yahwist-Elohist writings that reflected their Aaronid interests rather than the Levite interests of the Deuteronomists.[5] These writings reorganized time, established the seven-day week, and emphasized sacrifice and ritual purity.

These priestly writings were completed after the fall of Babylon and the rise of the Persian Empire when the exiles had returned to Jerusalem. Through the Exile and the return, they served to acknowledge the end of the Davidic dynasty and legitimate the rise of the priestly rule in Israel, a rule in which the high priest assumed the power and privilege of the king.

This time of return in the Persian period brought forth a great flowering of biblical literature. All the prophetic literature, apart from Daniel, was given its final form. All the wisdom literature, except Ecclesiasticus and the Wisdom of Solomon, was completed, and all the historical books were finished, except for Judith and the books of Maccabees, which had yet to be lived. These remaining books came into being after the fourth century B.C.E. conquests of Alexander the Great and the introduction of Hellenistic culture in the lands of the ancient Near East.

In following this brief sweep of history, it becomes evident that the Hebrew Scriptures emerged in an evolutionary way as the reflections and writings of a community of faith. Consequently, what we find when we read the Scriptures is not just a chronicle of history, written by the powerful or the ones who survived, but the refined expression of religious experience and theological reflection within a community of faith. It is the work of the community itself.

When we reflect on this process, it may seem that this continuing interpretation and reappropriation of the Scriptures invalidates them in some way or deprives them of their claim to inspiration. This is not true. Reflection, questioning, evolution of thought, and reappropriation for contemporary times is not only characteristic of, but required for, all reflective self-awareness and growth.

In writing on the interpretive task, Wisdom scholar Roland Murphy addresses these issues:

> It is increasingly felt that the scholarly task has to engage itself on both sides of meaning, and not merely confine itself to the historical meaning of the past. Moreover, the interesting issue was raised: How many meanings does an ancient text have? Is the historical meaning of the past the only valid meaning? Is there a plus meaning inherent in the ancient text that is reached by the modern reader in a valid way? These questions receive an affirmative answer in the history of interpretation of the Bible. It is a matter of historical fact that interpreters of the Bible have consistently gone beyond the past historical meaning of a biblical text to find a meaning for their own day.
>
> (*Wisdom Literature and the Psalms,* p. 44)

Murphy continues, applying this continuing reality to each of us in our day:

> How does one come to the meaning the ancient text has for today? The posing of this question flows from two undeniable facts. First, every piece of literature has an afterlife. The meaning it had for its audience and generation yields to the further meanings it carries for successive generations who read it in a new light. This is true of any literature, the Bible included. (*Wisdom Literature and the Psalms,* p. 45)

The Christian Scriptures

The Christian Scriptures emerged in much the same way as the Hebrew Scriptures. Just as the earlier Scriptures had begun with the Mosaic Exodus and the Covenant with Yahweh on Mount Sinai, the Christian Scriptures began with the Kerygma or Good News announcing the Resurrection or passing over of Jesus and the new and everlasting covenant with the God of Israel.

Over time, the stories of Jesus, his teachings, his parables, and his miracles, were added to this central theological proclamation, and a protogospel of the sayings of Jesus, known as Q, *Quelle* or Source, came into being.

The writers of the synoptic gospels of Matthew, Luke, and possibly Mark built on these sources, using additional material and developing their own perspectives. The Gospel of Mark, which is generally assumed to be the first gospel to be completed probably addressed to a community of Gentile Christians. Looking to the unique concerns of the community, this abrupt and almost breathless narrative centers on Jesus as the Son of God, the Messiah or *Christos*. It portrays his rejection by many of his own people and his acceptance by at least some of the Gentiles.

The later gospels of Matthew and Luke draw upon both *Quelle* and the Gospel of Mark. These gospels were also written as pastoral documents and were directed to the unique needs and backgrounds of particular communities. Matthew directed his writing to Jewish Christians who were steeped in the Hebrew tradition. He proclaimed Jesus as the fulfillment of the law and the prophets. Luke, the physician who wrote for a Gentile community, communicated a special compassion and sensitivity for women and for the poor.

The first Christian Scriptures to be written, however, were not the synoptic gospels but the letters of Paul. Letters were the primary means of communication with distant peoples in this time and were well used in the early Christian communities. Most often, a letter simply made the rounds within the churches, being passed along from community to community. This practice, initiated by Paul, continued within the church throughout the entire New Testament period.

The Gospel of John was one of the last canonical documents to be written. Attributed to a disciple in the Johannine community at Ephesus, it portrays and speaks the words of the resurrected Jesus, the preexistent Eternal One who has become flesh to make the Father known.

The final book in the biblical canon is the Revelation of John. Written in the genre of the Jewish apocalyptic, which began with the visions of Ezekiel and was very popular in the Jewish community during the first century, it carries a decidedly dualistic perspective, looking to the destruction of the old and imperfect order and the coming of a radically new reign of God.

Many of these Scriptures were carried together by the Christian and Jewish communities through the first century until their traditions diverged because of political sanctions imposed by their Roman rulers. From this time on, the Jewish communities looked to the authority of the Torah as taught by their rabbis, and the Christian communities looked to the authority of the apostles expressed through their bishops and priests. Within the Christian community, these Judeo-Christian Scriptures circulated freely with other gospels, letters, and writings until the fourth-century reign of Constantine, when the present collection was declared to be inspired and the canon of Scripture was officially closed.

To say that there are no longer inspired writings is to discount the unnumbered men and women of God who have continued to speak and write, to teach, prophesy, and communicate wisdom, in the two thousand years that have passed since the last of the canonical writings. Yet this earlier body of writing has become normative and formative for the Christian community and finds its home in the depths of our hearts and the marrow of our bones. It forms our thoughts, our actions, and our way of being in the world. It is and will always remain the faith history of our beginnings.

Part One

THE HEBREW SCRIPTURES

Chapter One

THE PREHISTORY OF GENESIS

From time immemorial, people of all cultures have gathered around their tribal fires to share their Creation stories in myth and ritual, in song and dance. Reflecting unique environments and histories, these stories vary from culture to culture yet seem to arise from the collective depths of human consciousness and answer deep questions of meaning and existence. In their primordial wisdom, they express profound yearnings for healing and transcendence and reconcile us to our God or gods and to the worlds in which we live.

The Judeo-Christian tradition honors these cultural memories in the prehistory of Genesis 1–11 and expresses our earliest understandings of what it means to be human and to inhabit the earth. While each story has an integrity of its own and arises from different social, political, and religious spheres, these stories need to be read together as a unified theological proclamation. Their unity is evident in that they have been gathered and edited by the priestly writer(s) and are held together by the literary device of semitic enclosure in which the beginning and the end are similar in a parenthetical way. Most important, they move together as human and covenantal history from our earliest beginnings and the growing violence and alienation among human beings to the subsequent ecological crisis of the Flood and God's universal Covenant of Peace with the earth and the beings of the earth.

As the foundational stories of our religious heritage, these stories contain our cosmologies or worldviews. They give meaning and order and purpose to our lives and underlie our cultures, our religions, our laws, and our lifestyles. They are the lenses through which we perceive and live out our reality. In our present day, these stories form the spiritual foundation for our ecological reflections. They help us to find our place in the universe, to recover the sense of the sacred in our world, and to begin the great and holy task of reinhabiting the earth.

The Yahwist Story

The earliest of these stories is that of the Yahwist in Genesis 2:4b–3:24. In this story, a man is created first, formed by a potter God from the clay of the red earth. A garden is then created for his delight, and animals are created to be his companions and friends. Last, a woman is formed to enter into an intimate partnership with the man. When this creation is completed, the human ones are entrusted with the cultivation and care of the garden, an entrustment that calls them to share in the creativity and responsibility of the Creator. This primordial habitat or ecosystem (*oikos* = house) reflects our earliest understandings of the role of humans in relation to God and to the earth. Walter Brueggemann describes this role most succinctly: "The destiny of human creation is to live in God's world, with God's other creatures, on God's terms" (*Interpretation — Genesis,* p. 40).

> At the time when the Lord God made the earth and the heavens — while as yet there was no field shrub on earth and no grass of the field had sprouted, for the Lord God had sent no rain upon the earth and there was no man [*adam* = of the ground or taken out of the red earth] to till the soil [*adama* = ground or red earth], but a stream was welling up out of the earth and was watering all the surface of the ground — the Lord God formed man [*adam*] out of the clay of the ground [*adama*] and blew into his nostrils the breath of life, and so man became a living being. (Gn 2:4b–7)

This story comes to us through the oral tradition of an early seminomadic people who established permanent agricultural settlements on the arable land in southern Palestine. It was a harsh and thorny existence for them, and they had to work carefully on the land to eke out their existence. Within this context, the story answers questions such as "Where did we come from? Why are we here? Why is life so hard?" It remembers an earlier time in the Tigris-Euphrates Valley when the land was rich and green and life was easier.[1] It tells us that God made us from the earth with the same care and concern that we give to the making of our pottery or artifacts, and that we are here to till and keep, to care for and cultivate, the land.

Archeological evidence supports this imagery. Michael Wood tells us that four thousand years ago, at the end of the third millennium B.C.E., the rich lands of the Tigris-Euphrates, the moist green setting of the Garden of Eden, were destroyed through the increasing needs of growing cities, overgrazing, poor agricultural practices, shifting riverbeds, and tribal wars. Wood suggests that Abraham and Sarah, our ancestors in faith, may, in fact, have come up from Ur as environmental refugees.[2]

These settings and cultural memories are reflected in our emergence "from the red earth," in the richness and diversity of the garden, in the need for the human ones to cultivate and care for the earth, and in the eventual affliction brought upon the earth through human alienation and violence.

This early creation story was carried through a paradigm shift from pastoral living to life in the royal court and committed to a written form in the ninth century B.C.E. during the opulent reign of Solomon. This was a time of extravagance in the royal court when Solomon engaged wisdom scholars from Egypt and Mesopotamia, acquired hundreds of wives and concubines, kept thousands of horses and chariots, gave cities away as gifts in return for political favors, and devastated the cedars of Lebanon with enforced laborers to supply timber for his many building programs (1 Kings 1–11). Responding to these lavish times, the Creation story becomes countercultural and calls the people to simplicity and humility, to the awareness that they are formed of the clay of the earth and entrusted with its care and cultivation. It tells them there are limitations to human wisdom and autonomy and invites them to return to covenant fidelity and live trustfully in the presence of God.[3]

This story is about an immanent God, a God who labors in creating man and woman, who walks and talks with these human creatures in the garden, who later rescues them out of oppression in Egypt, who travels with them in the desert, and who is present to them in all their afflictions.

The Priestly Account

The later Creation story of Genesis 1:1–2:4 was authored by the priestly writer(s) who gathered these ancestral stories in the sixth century B.C.E., and is addressed to a community of despairing exiles who had remained faithful during their time of captivity in Babylonia. This was also a time of shifting paradigms. The people had lost their land, their temple, and their king. They needed to see things in a new way. They needed a new vision that fit their experience, that sustained them in the present moment and carried them into the future.

Responding to this time of grief and humiliation, the priestly writer draws upon the creation stories and cosmologies of Egypt and Mesopotamia and goes beyond them with the bold theological proclamation that God has called the universe into being, that God can be trusted in all circumstances, and that despite all oppression, poverty, and despair, humans are created in the image of God and are beings of dignity and honor.[4]

In this story, the habitable earth and all that exists are created out of a chaotic nothingness — a dark, watery abyss — through the movement of a mighty wind and the sovereign Word of God. Dry land is brought forth to make a habitable earth for plants, animals, and humans — for all living beings.[5] Humans, made in the image and likeness of God, are commissioned as God's representatives, called to relationship and responsibility in this well-ordered world. Finally, having continually affirmed its goodness, God rests and takes sabbatical delight in this great work of creation.

> Then God said, "Let us make humankind in our image, according to our likeness; and let them have dominion over the fish of the sea, and over the birds of the air, and over the cattle, and over all the wild animals of the earth, and over every creeping thing that creeps upon the earth."
>
> So God created humankind in his image,
> in the image of God he created them;
> male and female he created them.
>
> God blessed them, and God said to them, "Be fruitful and multiply, and fill the earth and subdue it; and have dominion over the fish of the sea and over the birds of the air and over every living thing that moves upon the earth."
>
> (Gn 1:26–28 [NRSV])

A Covenant of Peace

This later Creation account is echoed in the re-Creation story of the flood. In this ancestral story, the habitable earth and all the living creatures of the earth have become dis-eased through the violence of humans. God's flood — God's torrent of grief and anger over the recalcitrant humans and their spoiling of the earth — threatens "to return the earth to watery chaos."[6] Yet God remembers Noah. God gathers the family of Noah and their companion animals in the interim habitat of the ark to dwell together in safety in the midst of the chaotic waters. Then, once again, God sends a wind to blow over the waters, and dry land and vegetation appear. Once more, all living beings, both humans and animals, are blessed to be fertile and multiply, and humans are reaffirmed as created in the image and likeness of God.

After this re-creation, in a unilateral act of disarmament, God, who has met violence with violence, hangs a warrior's bow in the sky, transforming it into a rainbow as a sign of God's unilateral and everlasting covenant of peace with the earth and all the beings of the earth:[7]

> This is the sign that I am giving for all ages to come, of the covenant between me and you and every living creature with you: I set my bow in the clouds to serve as a sign of the covenant between me and the earth. (Gn 9:12–13)

ECO-SPIRITUAL REFLECTIONS

Several important themes arise in these primordial stories and flow through the entire Judeo-Christian Scriptures. These themes intermingle in such a way that they cannot be separated from one another, but we can identify them and reflect on them.

The Sovereignty of God and the Unity of Creation

The first of these themes is the sovereignty of God and the unity of creation. It arises in the priestly account and differs from other ancient Near East Creation stories in that the universe comes into being not through a great battle of the ancient gods or the mating of primordial gods but through the sovereign Word of Yahweh, who pushes back the chaos to create the heavens, the earth, and the beings of the earth. This sovereignty expresses the autonomy and initiative of Yahweh, the unity and goodness of all creation, and the absolute dependence of all creation on Yahweh.

This Creation story reflects the prescientific Hebrew cosmology in which the great protective vault of the sky covers the earth, holding back the chaotic waters above and allowing the waters below to recede and the dry land to appear. In this cosmology, the sun, moon, and stars hover below the vault, and floodgates in the vault allow the rain and snow to fall. The earth itself rests above the sea on columns that rise from its watery depths. In this imagery, watery chaos surrounds and forms the perimeter of a static and firmly fixed cosmos.

Heavenly Seat of the Divinity

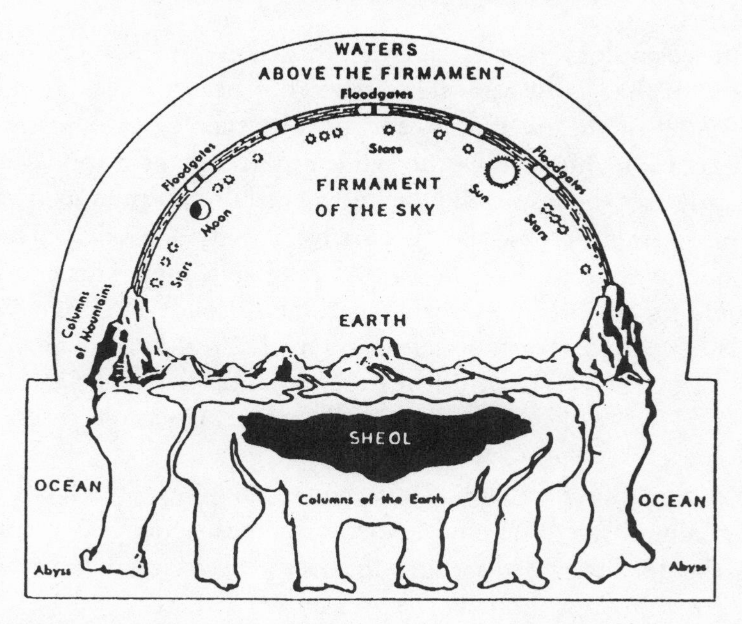

Reproduced with permission from St. Joseph Edition of the New American Bible.

While this cosmology is no longer valid in our scientific world, it needs to be understood and respected within the context of its time. Ted Peters writes:

> It is only to be expected that the biblical writers would reflect the cultural milieu and world view of their respective context and not that of a scientific age some 2,500 years into their future. Our own situation is parallel. Our cosmological thinking today cannot help reflecting the world view of mod-

> ern science, which provides the framework within which we do our thinking.
>
> ("Cosmos as Creation," *Cosmos as Creation,* p. 58)

German theologian Wolfhart Pannenberg not only echoes this wisdom, but also urges us to use it as a model:

> The cosmology that comes to expression in this idea of a vault of heaven is very impressive, but need not oblige the believer of the twentieth century. The theological doctrine of creation should take the biblical narrative as a model in that it uses the best available knowledge of nature in its own time in order to describe the creative activity of God.... This model would not be followed, if theology simply stuck to a standard of information that became obsolete long ago by further progress of experience and methodical knowledge.
>
> ("The Doctrine of Creation and Modern Science," *Cosmos as Creation,* p. 172)

Although we live with a different cosmology — which may be equally obsolete within our own lifetimes — many contemporary theologians are searching for areas of consonance between the Creation story and the new cosmological story we have received from our scientific community. Although they are different from one another in many ways, these stories resonate well in several areas. First, there is a profound ordering in the Creation account and in the known structure of the universe. Second, in both stories, creation emerges in sequential days or epochs. Third, this emergence begins with primal energy and moves through the emergence of inanimate matter and simple life forms to life forms that are far more complex.

When we reflect on this consonance in the primacy of light, for example, contemporary astrophysicists tell us that photons that carry the electromagnetic energy of light were among the first particles formed in the creation of the universe and that electromagnetic energy is one of the four great forces of the universe.

Both of these came into being in the first micromoments of the universe. We also know that life first emerged in the seas, in single-celled organisms, and that land animals and humans were the last to appear on the scene.

In reflecting on these consonances, Ted Peters suggests that the language of science and the language of theology might illuminate one another, and that while the languages differ, they are speaking of the same reality:

> Hypothetical consonance . . . recognizes where we are at the outset: theology sings one song and science another. Nevertheless, it gives us something to listen for. We can listen for those measures where both make sound at the same frequency, where we hear a momentary bar of harmony. Then we can at least ask if this might some day lead to a shared melody. ("Cosmos as Creation," *Cosmos as Creation,* p. 16)

In celebrating creation, we often give praise for the more tangible and objective realities we see around us, for plants, animals, wind, rain, and fire, but we would do well to remember that God also created the processes of the universe. God created time, gravity, the curvature of space, and space itself. God even created creativity. We are coming to understand that ongoing creativity is structured into the order of the universe. Through an element of possibility and indeterminance akin to free will, creatures and systems continually self-organize, self-actualize, and transcend themselves through their own creativity, which images and participates in the creativity of God. Trees bud forth, flower, and bear fruit in active participation in and with the creativity of God.

Nonhuman Creation

A second theme is the inherent goodness of creation and the value of nonhuman creation apart from the human. In the priestly account, we see Yahweh continually reflecting on the goodness of creation long before the coming of humans, and scholars are quick to point out that we do not arrive on the scene until the sixth day. Even when we do arrive, we share our day of creation with the land animals.

This inherent value of all creation, and its preciousness in the sight of God, is celebrated in the Psalms (104, 147), in the Book of Job (38–39) and in the Book of Daniel, where Azariah and his companions walk about in the fire, glorifying God and inviting all creation to join them in praise:

> Bless the Lord, all you works of the Lord,
> praise and exalt him above all forever.
> Angels of the Lord, bless the Lord,
> praise and exalt him above all forever.
> You heavens, bless the Lord,
> praise and exalt him above all forever.
> All you waters above the heavens, bless the Lord;
> praise and exalt him above all forever....
> Fire and heat, bless the Lord;
> praise and exalt him above all forever.
> Frost and chill, bless the Lord;
> praise and exalt him above all forever....
> All you birds of the air, bless the Lord;
> praise and exalt him above all forever.
> All you beasts, wild and tame, bless the Lord;
> praise and exalt him above all forever....
> Spirits and souls of the just, bless the Lord;
> praise and exalt him above all forever.
> Holy men of humble heart, bless the Lord;
> praise and exalt him above all forever....
>
> (Dan 3:57–87)

We also find nonhuman creation as teacher in the Book of Job (12:7–9), and Jesus calls us to learn from it in the parables and sermons of the gospels (Mt 6:25–34; 13; Mk 4; Lk 6:43–45; 8:4–18; Jn 10:1–18; 15:1–11). Even the Book of Revelation honors the nonhuman world in the animals that surround the heavenly throne. These animals are the beings that are closest to God and lead all the others in worship and praise (4:1–11).

Through these Scriptures we learn that it is not only the human who gives praise to God. The indwelling Spirit lifts all creation in praise. These proclamations also remind us that the natural world is not ours to abuse or exploit. The whole world, indeed the whole universe, is God's creation and is to be celebrated, cherished, and gratefully received. As Christians, our first call, therefore, is to radical amazement. Our second is to gratitude. We need to open ourselves to wonder and awe, to give ourselves over to worship and praise, to reflect back to God this Holy One's own goodness and glory.

In our communal life of faith, we have come to believe that God creates the world out of freedom and love, that the world is intended and willed, that it has a beginning and a final fullness. We believe that through the Word and the Spirit creation unfolds, continually transcends itself in time, and moves forward to completion. Through this evolutionary process, creation comes forth and returns to God without ever leaving God. Within this experience of creation, all things exist in God and God exists in all things. Because of this, God's presence to creation is always both immanent and transcendent.

In God's creation, everything is connected to everything else. Created and contained in the Word, ensouled and bonded in the Spirit, creation is a community. It is a single unified whole where everything is sacred. Through this unity of creation, we come to a deeper understanding of God as sacred community, of God as Creator, Redeemer, and Sanctifier. Creation as sacred community reflects this inner life of the sacred community of God. It, too, is relational and participatory. It is one and it is many, There is unity and there is diversity. Because of this, it is not only humans who image God, but the whole of creation. Saint Thomas Aquinas tells us that the whole of creation reflects and is needed to reflect the Divine Image.[8] All things are inherently good because they reflect the goodness of God.

It is urgent that we be mindful of this in our day. With the extinction of species and the degradation of land and sea and air, God's creation is diminished and many reflections of God are extinguished forever. With this diminishment and loss, our psycho-spiritual lives, our cultural heritages, and our sacramental rituals lose the richness and purity of their symbols and meanings.[9] Beyond this, with the loss of beauty and diversity on the planet, the human spirit itself is diminished, and the magnificent chorus of praise is quelled.

Human Ecology

A third theme is that of human ecology, a theme we usually reduce to a human supremacy or "dominion," forgetting our immersion in the community of creation and our creaturely and priestly call to appreciation and worship, to wonder, awe, and praise. In our human arrogance, we have often thought of ourselves as being above nature, but we are truly earthlings, a part of God's creative venture. We are living beings among other living beings. We are born and we die. We breathe the same air, require the same water, and share the same habitats. For better or for worse, we are part of the ecosystems in which we live.

Scientists remind us that in our humanity we image the unfolding of creation. Our bodies are about two thirds seawater, our fetal development recapitulates the development of less complex life-forms, our primitive brain is reptilian, and we most probably learned our speech through imitating the sounds of wind, rain, birdsong, and land animals. In perceiving us as creatures in the fellowship of creation, Jürgen Moltmann writes:

> ... before we interpret this being as *imago Dei,* we shall see him as imago mundi — as a microcosm in which all previous creatures are to be found again, a being that can only exist in community with all other created beings and which can only understand itself in that community.... Moreover as the last thing to be created the human being is also dependent on all

> the others. Without them his existence would not be possible. So while they are a preparation for him, he is dependent on them. (*God in Creation,* pp. 186–87)

Moltmann also tells us that in this existential stance of *imago mundi,* human beings are the embodiment of all other creatures and stand before God as their priestly representatives.

> Understood as *imago mundi,* human beings are priestly creations and eucharistic beings. They intercede before God for the community of creation. Understood as *imago Dei,* human beings are at the same time God's proxy in the community of creation. They represent his glory and his will. They intercede for God before the community of creation. In this sense they are God's representatives on earth.... Human beings are at once *imago mundi* and *imago Dei.* (*God in Creation,* p. 190)

It is this *imago Dei,* this representation of God to the community of creation, that is the great issue of our day. We know quite well that the Creation stories commission us to care for creation, but we still do not quite understand who we are or the meaning of this commission.

The priestly account commissions us in the more regal and masculine language of *subdue* and *dominion.* The Yahwist story uses the more pastoral and feminine language of *cultivate* and *care.* Those in power have favored the regal and masculine language and often understood it as a supremacy that entitles us to use and even abuse the earth and the creatures of the earth. It needs, rather, to be understood as a commission to foster their well-being by mediating the just and compassionate rule of God in this small corner of the universe. The biblical understanding of *dominion* is not a command to "lord it over" creation but a call to relationship, responsibility, and service.

In tracing the roots of this commission, we find that in the creation myths of Babylonia, the gods created humans to do the work of the gods so the gods could rest. When these gods had finished creation, they retired to the sanctuaries that had been built for them and rested or were enthroned. These gods were now transcendent and remote. Above and beyond creation, they entrusted the responsibility for the earth to humans.

Appropriating this perception and moving beyond it, the priestly writer(s) envisions a sovereign and transcendent God who is Lord of Creation. He perceives humans imaging God in creation through the work that they do and names this human responsibility in the words *rada/radah* and *kabash/kavash*.

As we all know, languages evolve and vocabularies grow. Biblical Hebrew contains only a few thousand words, and the original texts have no vowel points. Because of this, each word has shades of meaning that arise within its varying contexts and needs to be understood in this way. In addition, words with similar appearances but different meanings are difficult to distinguish from one another in old, damaged, or closely written manuscripts, and each word loses aspects of its meaning when it is translated from one language, culture, and/or age to another.

In the Hebrew language, *rada/radah* is understood as *to rule, subdue,* or *tread down* as one would tread down grapes in a winepress. In its earliest biblical setting, it expresses the ideal of kingship in Israel, which is patterned after David, the shepherd-king. Imaged and celebrated in Psalm 72, this ideal shepherd-king will govern the people with justice, defend the afflicted ones who cry to him, and rescue the lowly and the poor. The psalm tells us that "Justice shall flower in his days, and profound peace, till the moon be no more" (Psalm 72:7).

In this setting, *rada/radah* is about intimacy and interrelatedness.[10] The ruling and subduing does not carry an aggressive element, but rather the sense of bringing creation into a state of peace as one would care for and subdue a beloved child waking from a bad dream. In the light of the priestly Creation story being a companion piece to the earlier Creation story and its use in Psalm 72, *rada/radah* explicates a positive rule of compassionate shepherding where justice, safety, and peace abound.

Kabash/kavash may also be understood as *subdue* and *tread down.* It relates to similar ancient Near East words meaning *tread down, knead,* and *press.* Translated as *subdue,* it carries a range of meanings from treading or trampling on one's enemies to the simple treading and trampling of the earth as one tills the soil and covers the seed. In its context as a companion piece to the Creation story about tilling and keeping, and in the context of the pristine creation where there are as yet no enemies or predators, the simple act of farming or tending the garden is the favored contextual understanding.[11]

This image of dominion as compassionate shepherding can be found throughout the Scriptures. We see it in the fidelity and service of Noah in sheltering the animals in the ark, a fidelity and service that led to a re-creation of the world and a covenant of peace with all creation. We see it in the just shepherding of the Lord in Ezekiel (34), the self-emptying of Jesus in the Philippian hymn (Phil 2:5–11), and the good shepherding of Jesus, who lays down his life for his sheep (Jn 10:10–11).

In the Ezekiel passage, Yahweh speaks to the inauthentic shepherds of Israel, proclaiming the authentic and compassionate rule of God:

> I myself will look after and tend my sheep. . . . I myself will pasture my sheep; I myself will give them rest, says the Lord GOD. The lost I will seek out, the strayed I will bring back, the injured I will bind up, the sick I will heal . . . I will make a covenant of peace with them, and rid the country of ravenous beasts, that they may dwell securely in the desert and sleep in the forests. (Ez 34:11–25)

We see this expression of authentic dominion today in the martyrs of El Salvador who have given their lives in the service of justice and the poor; in the Chipko women of India who defend their forests with their own bodies; in the men and women who have nursed the American bison and the California gray whale back from near extinction; in the political prisoners interred for their protests against unjust political systems or the destructiveness of nuclear weapons. This is true dominion in the image of the Creator.

The Sabbath

A fourth theme is that of the Sabbath. In our anthropomorphism, we have often looked to ourselves as the crown of creation, but the crown of creation is the Sabbath. It is the feast of creation, the day of rest and contemplation when God, like a great artist or artisan who has completed a great work, reflects on it, both letting it be in itself and taking it back into the Divine Self in a new way. In this day of Sabbath rest, God knows and embraces and blesses creation, rejoicing in the beauty and order of what has been brought into being.[12]

This Sabbath also allows creation to come into its own, to know itself in the presence of God. It allows creation to unfold, to come into history and move toward its inherent goal. Jürgen Moltmann writes:

> In his present rest all created beings come to themselves and unfold their own proper quality. In his rest they all acquire their essential liberty. By "resting" from his creative and formative activity, he allows the beings he has created, each in its own way to act on him.... But on the Sabbath the resting God begins to "experience" the beings he has created.... He allows himself to be affected, to be touched by each of his creatures. He adopts the community of creation as his own milieu. (*God in Creation,* p. 279)

In the Sabbath of the Judeo-Christian Scriptures, God rests in the sanctuary of creation but also in the sanctuary of divine solitude beyond creation. God is enthroned in the midst of creation and remains sovereign beyond creation. In this way, through the Sabbath, God exists and is made known in both immanence and transcendence, in time and in timelessness. This Sabbath and all the Sabbaths that follow celebrate these beginnings and prefigure the cosmic Sabbath of end-time when creation will come into its final fullness.

The Jewish Sabbath may have had its origins in prehistory, in tribal rituals around work and leisure, or as a response to the Hebrews having no rest as slaves in Egypt. We do know that the Sabbath became a matter of law among the tribes of Moses through their encounter with Yahweh on Mount Sinai, and its *shalom* was to be extended not only to strangers and guests but to the animals and the land.

This early Sabbath came into its full flowering during the time of exile. The prophet Jeremiah had warned the people that their failure to keep the Sabbath would bring destruction, and when destruction came, they remembered. Without a temple or a land, the Sabbath became not only a time but also a place of meeting God.[13] It became a time and a place of rest and contemplation, a time and a place to sing their songs, to tell their stories, and to dream their dreams of a new beginning.

In the Christian tradition, the sacred character of the Sabbath has passed over into Sunday. The history of this movement is obscure and may have involved both practical and political issues, but we have come to understand it as the messianic extension of the Jewish Sabbath, which has come into being through the death and resurrection of Jesus.[14] In the Hebrew Scriptures, the first Sabbath celebrates the completion of creation when God rests in creation and returns and rests in the wholeness of divine solitude. In a similar way, in the death and resurrection of Jesus the initial work of the new creation has been completed and Jesus returns to a restful enthronement in the heavens, sending his Spirit to rest in the new creation.

Jürgen Moltmann writes of the Sabbath as a heritage and an image of liberation:

> Israel has given the nations two archetypal images of liberation: the exodus and the sabbath. The exodus from slavery into the land of liberty is the symbol of external freedom; it is efficacious, operative. The sabbath is the symbol of inner

> liberty; it is rest and quietude. The exodus is the elemental experience of God's history. The sabbath is the elemental experience of God's creation. The exodus is the elemental experience of the God who acts. The sabbath is the elemental experience of the God who is, and is present. No political, social and economic exodus from oppression, degradation and exploitation really leads to the liberty of a humane world without the sabbath, without the relinquishment of all works, without the serenity that finds rest in the presence of God. (*God in Creation,* p. 287)

Sabbath/*Shabat* literally means *rest.* Most of us do not really understand this rest. We live in a restless society and a restless world. We do not allow ourselves or our children to rest. We do not even allow our machines and our technology to rest. Our recreation itself is restless and aggressive, and tourism and theme parks replace pilgrimages to sacred places of rest. Like Augustine's, our hearts are restless, and whether we listen to them or not, they cry out for a resting place in creation and in God.[15]

In the Gospel of Matthew, Jesus speaks of this rest to the poor of the land, to those who are wearied and burdened by the yoke of religious legalism and political oppression:

> Come to me, all you who labor and are burdened, and I will give you rest. Take my yoke upon you and learn from me, for I am meek and humble of heart; and you will find rest for yourselves. For my yoke is easy and my burden is light.
>
> (Mt 11:28–30)

In this passage, Jesus invites the restless and world-weary to enter into and participate in God's own rest. He tells us that this rest, this inner peace, is found through meekness and humility of heart, through a strong gentleness or gentle strength that arises from integrity and inner authenticity and manifests itself in nonviolence and compassion within the community of creation. He tells us that in patterning our lives on his own, we come into and abide in this rest.

To rest in God, then, is to abide, here and now, in the timeless being of God. It is to let go of our restless strivings and to live and move in the world in a compassionate and nonviolent lifestyle that draws its wisdom and power and will from the wisdom and power and will of God. In its deepest reality, it incarnates the nonviolence and compassion of God made manifest in the life of Jesus and enables us to participate consciously and fully in the work of the new creation. It becomes an eschatological call to root and ground our lives in God and in God's will, and live and work toward the final fullness of the reign of God.

Covenant

Throughout the Scriptures, God reaches out to creation in "caring" and "cultivates" relationship with the earth and the beings of the earth through Covenant. The first Covenant is generally understood to be the blessing of all creation on the first Sabbath. The second is the Covenant with Noah, the earth, and all living beings. The third is the covenant with Abraham. In the Hebrew Scriptures, the great covenant is the Sinai covenant with the tribes of Moses, and in the Christian Scriptures we find a new and everlasting covenant of peace with all creation in the death and resurrection of the Lord Jesus Christ. This continual reaching out to creation through covenant manifests the compassionate "dominion" and sovereign "rule" of Yahweh in creation.

Covenant, as a concept and as an experience, is as ancient as human existence, and expresses the enduring need to establish patterns of relationship between individual persons and among diverse peoples. In the ancient Near East, covenants were usually initiated with solemn oaths, shared meals, exchanges of gifts, and the summoning of witnesses such as mountains, rivers, stones, and gods. In entering a covenant, persons or tribes were bound to one another as blood brothers or sisters and acquired duties and responsibilities that sustained the relationship. This concept was well understood in the ancient Near East where individual persons or warring nations frequently established treaties and covenants to avoid or mitigate conflict or the violence of war.

These covenants followed two basic forms. The simplest of these was a treaty or formal agreement between two mutually concerned parties, or the formal promise of blessing or reward given by a ruler to a loyal servant. The covenants of Genesis reflect these simple covenants. In the first Creation story, God simply blesses creation and asks nothing of it in return. In the Noah story, God gives a blessing once again, and also gives the rainbow as a sign and a reminder of the covenant. Again, God asks nothing of creation. These covenants are unilateral and unconditional. The covenant with Abraham repeats this blessing in the gifts of a multitude of descendants and the promise of "the land," but requires circumcision as a reciprocal sign of the covenant.

The second and more complex form of covenant was the suzerainty treaty, which was predominant in the Hittite Empire during the time of Moses. In this treaty form, the greater power, or suzerain, established a treaty with a weaker, vassal tribe or nation and gave protection in exchange for allegiance and tribute, usually in the form of military service, enforced labor, grain, wine, and oil. If the lesser party rebelled, failing to give allegiance or tribute, or turned to another great power for protection, the original power would probably turn on the vassal state and destroy it.

The Mosaic covenant reflects this suzerainty treaty and the turmoil of its social, political, and historical time. Israel itself came into being among the Hebrews (*abiru* or dispossessed peoples) of Palestine who were rebelling against the existing order and its coercive forms of power. Too weak to stand alone, the tribes of Israel first banded together in a loosely knit six-tribe confederation and later joined the tribes of Moses in their covenant with Yahweh when the Moses tribes came into the land of Israel. In describing the origins of these Hebrew people, Robert Coote writes:

> The one characteristic that applied to all *abiru* was that they had been uprooted from one political and social con-

> text and forced to adapt to another. An *abiru* was a migrant or displaced person in a new social location.... Reasons for breaking former ties include war, famine, lack of opportunity at home, personal disaster, debt, excessive taxes, and lengthy military service.... The people called *abiru* in different places and times had nothing more in common than their transitional social condition. (*Early Israel,* pp. 42–43)

While the Hebrews had a reputation for being warlike, they were also war-weary. As the socially and politically oppressed, they lived deeply amid their experiences of war and their dreams of peace. They were tired of slavery and oppression and longed for a time and a place of peace where each one could sit under their own vine or fig tree, where they could live in the houses they had built and eat the fruit of the vineyards they had planted (Isaiah 65:21). As an expression of their objection to the incessant wars, they invested Yahweh with the decisions of making war and establishing peace. As an expression of their objection to the oppression of local rulers and strongmen, they looked to Yahweh as their only Sovereign Lord and King.

Through their interaction with neighboring nations, the Israelites came to understand their covenant with Yahweh as both similar and dissimilar to the treaty of the Hittites. It was similar in that Yahweh was a greater power and would allow or bring punishment upon the people if they were unfaithful or violated the laws of the covenant. It was dissimilar in that Yahweh would never abandon the people. Yahweh was always willing to take them back, calling them, again and again, to change their hearts, to remember, to repent, and to return.

In this way, the covenant was both conditional and unconditional. While Yahweh, through the aspect of justice, would allow disaster to befall the people, this Holy One, through the aspect of *hesed* or faithful love, would continue to seek them out, calling them to return and bringing back those who were lost or who had strayed.

These experiences of covenant speak of a God who extends

mercy and compassion, who is slow to anger and rich in mercy, whose capacity to love and suffer and forgive in the midst of rejection, abandonment, and betrayal is greater than our capacity to reject, abandon, and betray.

Through this enduring love, creation and covenant are inseparable. Creation expresses the wisdom and power of God in bringing all things into being. Covenant expresses the compassion and faithfulness of God in bringing all things into a final fullness. Covenant expresses God's willingness to suffer in and with creation, to remain faithful to creation, and to honor and embrace its freedom, its limitations, its anguish, and its failures.

In covenant, God takes into the Divine Self the nonbeing and nothingness of the world and transforms it into God's own life. In covenant, God indwells and gathers to the Divine Self the nonbeing and nothingness of all things — the lilies and the birds, the stones and the fields, time and history and the human. All things come forth, abide, and find their final fullness in God. We move toward this unfathomable mystery in creation and covenant.

Chapter Two

THE LAW

The Law of Israel is first and foremost about Covenant. It is a divine revelation that offers the saving gift of life within the relational context of Covenant and establishes the contours of all human relationships — relationships with God, among humans, with nonhuman creation, and with the land. For Israel, this law is not burdensome. Because it springs forth as saving grace, it is a source of meditation and a cause for joy. This delight in the law is celebrated throughout the Scriptures:

> The law of the Lord is perfect,
> refreshing the soul;
> The decree of the Lord is trustworthy,
> giving wisdom to the simple.
> The precepts of the Lord are right,
> rejoicing the heart;
> The commands of the Lord are clear,
> enlightening the eye;
>
> (Ps 19:8–9)

In the traditional theology of Covenant, the law does not reside in Yahweh as Creator but in Yahweh as Liberator. Yahweh is the Holy One who through the central Israelite experiences of Exodus and Sinai brings the people out of slavery and into freedom. Unique among all the gods, Yahweh is a God who acts in history.

This sovereignty and saving power of Yahweh is expressed in the one great law that surpasses all others:

> Hear, O Israel! The LORD is our God, the LORD alone! Therefore, you shall love the LORD, your God, with all your heart, and with all your soul, and with all your strength.
>
> (Dt 6:4)

Developing through a series of legal codes over a period of six hundred years, the law of Israel looks to Moses as the great lawgiver and flows forth from the community's understanding of his spirit. In its evolution, it reflects a growing awareness in social, political, and religious consciousness and forms a living heritage to guide the people and the nation. Throughout its development, it addresses everyday activities and community life and evolves and adapts to changing times and situations. Gerhard Von Rad writes:

> Israel regarded the will of Jahweh as extremely flexible, ever and again adapting itself to each situation where there had been religious, political, or economic change. . . . Jahweh's will for justice positively never stood absolutely above time for Israel, for every generation was summoned anew to hearken to it as valid for itself and to make it out for itself.
>
> (*Old Testament Theology,* p. 199)

The Codes of the Covenant

The earliest understanding of Yahweh's will comes to expression in the Decalogue or Ten Commandments written on the tablets of stone at Sinai (Exodus 20:2–17; Deuteronomy 5:6–21). This code of law comes through Moses and the Yahwist tradition of the Southern Kingdom in the tenth century B.C.E. The later Code of the Covenant, which is based on an earlier moral code, expresses concern for widows, orphans, aliens, and slaves as well as animals, the Sabbath, and the land (Exodus 20:22–23:33). This code addresses a settled community of herdsmen and agriculturists and finds its way into the Scriptures through the Elohist tradition of the Northern Kingdom in the eighth or ninth century B.C.E.

The Deuteronomic Code adapts the Code of the Covenant to changes in social and economic life and includes a concern for tithes, festivals, worship, and sacrifices, as well as the care of slaves and the remission of debt (Deuteronomy 12:1–26:15). This code comes through the Deuteronomic tradition of the Southern Kingdom in the seventh century B.C.E. and centralizes worship in the temple of Jerusalem. The final set of laws, the Holiness Code, is a collection of laws concerning ritual purity and the Sabbath (Leviticus 17–26). These laws were gathered and refined over time through the Levitical priesthood in Jerusalem and came into their final form through the priestly writer(s) during the time of exile and return.

Each of these codes perceives all aspects of life as existing within the community of Covenant. In perceiving the whole of the Israelite experience within the community of covenant, the law confers rights on the whole covenant landscape. It is a shared and sacred trust that embraces the land, the animals, the alien, and the poor. When covenant relationships are honored, prosperity abounds. When covenant relationships are violated, the whole landscape suffers. In this way, the covenant community is experienced as an interconnected and interdependent web of relationships. The violation of any strand damages the wholeness of the web and threatens to return creation to chaos. Quite simply, as our contemporary experience tells us, creation is held together and sustained through covenant.

ECO-SPIRITUAL REFLECTIONS

In looking to what we might call the environmental ethics of Israel, it is interesting to note that the laws regarding the land are usually related to the Sabbath and the love of God while the laws relating to animals are usually associated with the love of neighbor. The Scriptures declare that the earth is the Lord's and we are only tenants on the land, but the animals are our kin, our brothers and sisters in the realm of "living beings" and "all flesh."

A Sabbath for the Land

The Sabbath laws of Israel invite all of creation to enter into the "rest" of the Creator and to find healing and restoration in this rest.[1] Emerging from the experiences of life, they honor the rhythmic and mutually supportive experiences of being and doing, giving and receiving, reflecting and acting. They also understand that rest and reflection deepen understandings, enhance meanings, and prepare us for renewed community.

Within the community of Covenant, every seventh year is a Sabbath year when the land rests, debts are forgiven, slaves are set free, and the poor and the wild creatures may feast on the resting land:

> For six years you may sow your land and gather its produce. But in the seventh year you shall let the land lie untilled and unharvested, that the poor among you may eat of it and the beasts of the field may eat what the poor leave. So also shall you do in regard to your vineyard and your olive grove. For six days you may do your work, but on the seventh day you must rest, that your ox and your ass may also have rest, and that the son of your maidservant and the alien may be refreshed. (Ex 23:10–12)

The Jubilee Year, the year after seven times seven years, is the most high of the Sabbatical years. It is the time of liberation and social reorganization when the land itself is set free to rest and to be returned to ancestral families or reapportioned among the poor. Though it was held out more as an ideal than a reality, it was understood as a time when the whole covenant landscape was to be restored to the divine ordering of Israel's beginnings:

> The fiftieth year you shall make sacred by proclaiming liberty in the land for its inhabitants. It shall be a jubilee for you, when every one of you shall return to his own property, everyone to his own estate. (Lv 25:10)

There is little need to speak of our failure, as a people, to respect the land. We know it all too well. We do not honor it or live wisely on it. We do not allow it to rest. We overcultivate our soil, poisoning and depleting it through the excessive use of pesticides, chemical fertilizers, and monocropping. We overcut our forests, causing loss of habitat, erosion, the silting up of rivers, and damage to aquatic life and the native peoples whose livelihood depends upon these resources. We tear up the land and blow mountains apart in mining nonrenewable resources that we could conserve through recycling or the use of renewable resource technologies. We overgraze our land, turning pasture into wasteland. In many of these situations, we subsidize this destruction through government programs. The whole of our covenant landscape is endangered.

Christian ethicists are not always in agreement on the rightness of any particular response. Issues are complex and confusing, and solutions are difficult to achieve. There are legitimate needs, and differing perspectives.

An example of this is the logging conflict in the temperate rain forests of the Pacific Northwest. The people who support intensive logging see the trees as an economic resource that supports the economy and provides jobs for workers. For these workers, there is the legitimate and terribly immediate need for employment and support for their families. For many there is also the economic belief that the forests belong to whoever buys the rights to harvest them.

The deep ecologists have a different perspective. They see the forest as part of the covenant landscape, in union and in communion with all the beings of the earth, belonging to the future as much as to the past and the present. In their perspective, the forests are a beautiful and fragile part of the eco-spiritual web. They are to be used with restraint and preserved for their own sake, for the well-being of nonhuman nature and for the generations yet to be born.

In struggling for solutions, we often reduce this conflict to the spotted owl versus the loggers. It is not that simple. A forest contains a multitude of other life-forms. In this situation, the spotted owl has become the barometer for about 160 other species in the old growth forest. It is like the canary in the coal mine. When things are not right for the canary, they are not right anywhere in the mine. In a similar way, when things are not right for the spotted owl, things are not right anywhere in the old growth forest.

One argument made by the logging interests is that they are just harvesting the trees. It is not like mining the earth, they say, the forest will grow back. This is not true. A forest that is cleared will not grow back. A Douglas fir, which is prominent throughout the Pacific Northwest and Southwestern Canada, will grow for about eight hundred years before it dies and falls to the forest floor. It will then decay slowly for about 250 years, returning nutrients to the earth and supporting new growth in the forest. The life cycle of a redwood is even longer. There are many redwoods that are over two thousand years old still standing in the old growth forests. If these trees are cut down and taken from the forest, there are no nutrients returned to replenish the forest. In addition, a broken canopy endangers the forest itself.

There is also the fallacy that forests can be replaced by tree farms. This is a very limited perspective. The trees are not the forest. The trees may be the largest form of life in the forest, but they are not the forest. A forest is an ecosystem. It consists of all the different plants, animals, streams, and geological formations that are present, as well as the processes through which these life-forms and life-systems interact with one another. When a forest is cut, many of these other life-forms must move on or die. They simply cannot inhabit a tree farm.

Another aspect of the situation, one that is rarely mentioned, is that many jobs in the logging industry are lost because we ship raw, unprocessed logs to places like Japan. Japan has laws that prohibit logging in its own forests, but it imports vast quantities of timber from the temperate rain forests of the Pacific Northwest and the tropical rain forests of the Philippines and Southeast Asia. In the Philippines and Southeast Asia, native peoples lose their homes and lands, and many species are extinguished through this logging as well as through agribusiness. These destructive enterprises are often needed to make payments on international debts to lending agencies, such as the International Monetary Fund. Caught in this cycle of poverty and destruction, native peoples who have been displaced then slash and burn their way up the forested mountains or deeper into the jungle where a year or two of their farming depletes the fragile forest soil.[2]

In all these cases, few people are attentive to the reality that these forests are the lungs of the planet. Forests take in carbon dioxide and breathe out the oxygen. This is especially important when we are already experiencing an excess of greenhouse gases in the atmosphere. Forests also moderate temperature and cycle water. Even a limited loss of these forests causes dramatic environmental change on a global level.

In reflecting on these environmental issues, many Christians push them aside, believing our first duty is to the poor and not to the earth, but the earth has become poor and afflicted too, and these issues of environmental concern and human poverty are deeply intertwined. Poverty and environmental degradation build on one another and form a downward spiral that affects all things. Ultimately, social justice involves the whole earth community.

Another dimension to our valuing the land lies in its ability to reveal the Divine Presence, to foster psycho-spiritual healing and growth, and to nurture the human spirit. We all know that we experience the presence of God more intensely in the natural world, in wilderness and in wild things, in well-kept fields and forests and gardens. With the loss of beauty, diversity, and wildness on the planet, we lose a sense of the sacred. We lose our sources of wonder and awe and our images for symbol and meaning. With this loss, our language loses its richness, and our sacramental life loses the purity of its base in images and symbols from the natural world. Without this re-creative power of the land, we become less than ourselves.[3]

The Presbyterian Eco-Justice Task Force reminds us of this reality:

> As human creatures, integrally bound up with other forms of life... we draw strength, inspiration, enjoyment, and fulfillment from the nurture of our relationships in the realm of nature. We are impoverished if we do not know, emotionally as well as intellectually, our need for the land, and if we do not celebrate its variety, fecundity, and beauty, or if we do not experience the pleasures and responsibilities of interacting with, indeed communicating with, loving and being loved by, other sentient creatures.
>
> (*Keeping and Healing the Creation,* p. 38)

Companions and Friends

The laws of Israel concerning animals are addressed primarily to domesticated animals, to those that we have taken out of the wild and brought into culture as property and for service. There is an ambiguity to these laws, however, for while the people are urged to respect these animals and care for their needs in a humane manner, they also sacrifice a multitude of animals as tithes or in atonement for their sins.

It is difficult for us to understand the sacrificial slaughter of all these lambs and doves and young bulls, and the Scriptures themselves assert that God has no need of these sacrifices, yet we ourselves have our factory farms and sacrifice countless animals in our desire for meat. Unlike the Israelite people, who treated their sacrificial animals with respect, we are often cruel and treat our sacrificial animals without compassion. In many situations, we keep them closely confined from birth, fill them with antibiotics and growth hormones, and slaughter them in ways that are terrifying and painful to them. We violate the whole of their life cycle, and when we eat their flesh, we eat of their fear and their tortured lives and deaths. This same violence to animals occurs in the use of animals for medical or cosmetic experiments, and to a lesser degree in the animals we keep in zoos and marine exhibits.

We know we do not need to relate to animals in this way, and we know we do not need to eat meat, certainly not large quantities of it. We would, in fact, be healthier if we did not do so. We also know that the grain required to raise one large animal would feed many people and that our planet could support our human population in a better way if we ate lower on the food chain and did not raise so many animals for food.

We also know we are called to respect these animals as companions and friends. Their lives are woven together with ours. Their pain and joy are like our own. Saint Francis of Assisi rightly called them brothers and sisters,[4] and Chief Seattle warned us that without them we would die of loneliness.[5] We know the truth of this in our relationships with our household pets. There is a deep current of empathy and sympathy between us. Their eyes mirror our own souls and express a deep and profound inner life. Gary Kowalski writes of this inner life:

> They are not an entirely different order of creation, but like us they have rich and spacious interiors. They contain inner landscapes: desert places and lonely canyons, cliffs of madness and rivers of serene awareness that merge in tranquil seas. They . . . are not our property or chattel, therefore, but our peers and fellow travelers. Like us, they have their own likes and dislikes, fears and fixations. They have plans and purposes as important to them as our plans are to us. Animals not only have biologies; they have biographies. We can appreciate the lives of animals, but not appropriate them, for they have their own lives to lead.
>
> (*The Souls of Animals*, p. 107)

Strangers and Guests

The tribes of Moses came unto the land of Israel as political refugees fleeing an oppressive regime. They were welcomed by some of the inhabitants and fought great battles with others. To be refugees, strangers, and guests is not a new experience for any of us. We have all found grace and mercy in some "foreign land."

Both the Genesis story and archeological evidence suggest that Abraham and Sarah may have come up from Ur as environmental refugees. The Scriptures also tell us that later these ancestors went to Egypt because of a famine in Canaan, and that still later the sons of Jacob went to Egypt when another famine struck the land. Because of these early experiences of famine and the harshness of desert life, Israelite law embraced a courtesy and compassion for strangers and guests. Again and again the Scriptures call us to this remembrance:

> You shall not oppress an alien; you well know how it feels to be an alien, since you were once aliens yourselves in the land of Egypt. (Ex 23:9)

This treatment of aliens and refugees, strangers and guests, speaks not only to our everyday experiences in multicultural communities, but also to our contemporary issues around political and environmental refugees. In our day, vast numbers of people are being driven from their homelands because of political oppression, environmental degradation, natural disasters, the ravages of war, and multinational encroachment. In times past, there have been lands to absorb this migration. We, in the United States, absorbed many of the refugees from the Irish potato famine, and our western states absorbed many of the rural poor of the Dust Bowl tragedy. Now, however, there are few places left for settlement. We have filled most of the habitable earth and have devastated much of it. As our soils die from deforestation, erosion, and poisoning, as limited wars and multinational encroachment continue, and as the cycle of environmental degradation and poverty intensifies, we will experience a greater measure of refugee activity than ever before.

This is a difficult truth. In itself, it calls us to live lightly on the land, to care for our own, to open our hearts to wounds of the world, and to cry out to God in prayer. But these times are not just about disasters, about geological, political, and economic earthquakes. They are also about something new — about birthquakes.[6] What we are learning about the earth, the people of the earth, and the earth process has a deep spiritual significance. We are learning that our disequilibrium cries out for a new equilibrium, that our chaos cries out for a new order, and that our migrations, whether they are physical, psychological, or spiritual, cry out for a new homeland.

When Yahweh called the Israelites out of Egypt, it was not just a movement out of a land. Walter Brueggemann tells us it was also a call to leave behind the weary and outworn consciousness of the empire and the politics of oppression and exploitation, and to enter into an alternative consciousness of radical amazement and a politics of justice and compassion. God was always free to do a new thing. Brueggemann writes:

> The task of prophetic ministry is to nurture, nourish and evoke a consciousness and perception alternative to the culture around us.... The alternative consciousness to be nurtured, on the one hand, serves to criticize in dismantling the dominant consciousness.... On the other hand, that alternative consciousness to be nurtured serves to energize persons and communities by its promise of another time and situation toward which the community of faith may move.
>
> (*The Prophetic Imagination,* p. 13)

Many of us believe that we are experiencing a call to a new consciousness now, a call to move beyond an anthropomorphic and nationalistic consciousness of racism, sexism, and human arrogance into a consciousness that embraces the whole human community and the whole earth community in a politics of justice, compassion, and inclusion. This cultural transformation is a new exodus, a new passing over that will require all our courage, all our hope, and a radical return to the wellsprings of divine being. It is a call made by God, a call to choose life and to seek and to follow the "new thing" that God is doing.

Chapter Three

THE PROPHETS

Prophecy is the keeper of covenant. It is a voice that calls and recalls a people to faithfulness, that cries out again and again to turn and return to a first love, to a love that was there in the beginning.

Prophecy cries out with the passion and grief of God in the brokenness of a wounded world. It reveals a God who is with us and for us, who is intimately involved in all that is human.

Prophecy cries out against anything that diminishes the human. It proclaims glad tidings to the poor and healing to the brokenhearted. It announces new sight to the blind and liberty to captives. (Isaiah 61:1–3; Luke 4:1–19) Prophecy is about God breaking through in our brokenness. It is about being carried into a place of freedom that energizes and leads to victorious songs of praise.[1]

Walter Brueggemann tells us that prophecy begins in a place of darkness. It arises in a place of pain and grief with the recognition that the present structures of reality, whether social, political, or religious, are not working. It begins with a cry to a God who can change things, who can bring people into a new freedom and light and who can make all things new. It reveals a divine pathos that lures a people out of the brokenness of the past into the radical newness of a future filled with hope.[2]

Prophecy is about transformation. It bears a coherence with tradition yet reaches out beyond tradition. It moves persons and communities out of that which is old and outworn into that which is radical and new and filled with the eternal freedom of God. In its transformational movement, it carries the tension of past and future together in the present moment and binds time and eternity in the eternal now.

Prophecy is a language of the heart. It is a poetry and lyricism that stretches language into a transformational newness, that spins the gossamer threads of words and images across the chasm between the word of God and human speech, between the freedom of God and human bondage. Prophecy calls all who will hear to integrity and authenticity, to fidelity and devotion. It moves from a law engraved on tablets of stone to one that is written on hearts of flesh, from a wisdom that is outside and beyond to one that is within one's own innermost being and makes its home in the human heart.

A Concern of the Heart

It is this concern with the heart and this language of the heart that occupies the prophets. They are the ones who cry out, again and again, from their own hearts to the hearts of others. "Cleanse your heart.... Say in your hearts... Pour out your hearts... Rend your hearts... (Jeremiah 4:14; Baruch 6:5; Lamentations 2:19; Joel 2:13). Speaking from their own inner depths, from what they have seen and have been moved by, from the divine pathos they have experienced in their own hearts, the prophets reveal the very heart of God.[3]

Abraham Heschel reminds us that the prophets do not come to this task with ease, that they rebel and turn from the distinction and affliction of the task. They are, rather, seized by God and lavished with deep insight and understanding, a gift that brings estrangement and pain. Heschel writes of this incarnational pathos of the prophet:

> The prophet is a man who feels fiercely. God has thrust a burden upon his soul, and he is bowed and stunned at man's

> fierce greed. Frightful is the agony of man; no human voice can convey its full terror. Prophecy is the voice that God has lent to the silent agony, a voice to the plundered poor, to the profaned riches of the world. It is a form of living, a crossing point of God and man. God is raging in the prophet's words.
> (*The Prophets,* Vol. 1, p. 3)

Heschel also perceives the prophet as the iconoclast, the one who is sensitive to injustice and evil, who sings an octave too high for the rest of us to hear. He visions the prophet's work as a scream in the night while the rest of the world is comfortable and asleep[4]:

> Instead of dealing with the timeless issues of being and becoming, of matter and form, of definitions and demonstrations, he [the prophet] is thrown into orations about widows and orphans, about corruption of judges and affairs of the market place. Instead of showing us a way through the elegant mansions of the mind, the prophets take us to the slums. The world is a proud place, full of beauty, but the prophets are scandalized, and rave as if the whole world was a slum. . . . Indeed the sort of crimes and even the amount of delinquency that fill the prophets of Israel with dismay do not go beyond that which we regard as normal, as typical ingredients of social dynamics. To us a single act of injustice — cheating in business, exploitation of the poor — is slight; to the prophets, a disaster. To us injustice is injurious to the welfare of the people; to the prophets it is a deathblow to existence; to us, an episode; to them a catastrophe, a threat to the world. . . . (*The Prophets,* Vol. 1, pp. 1–2)

The Message of the Prophets

The message of the prophets arises from the tradition of Moses, the first of the prophets, and addresses three major themes — fidelity to Yahweh, social justice and a concern for the poor, and a future time of salvation. All of these themes are rooted in Israel's foundational experiences of exodus, covenant, and entrance into the Promised Land. They recall Israel's origins as a people who were poor and oppressed, who were rescued and cared for through the mighty works of Yahweh, and who were brought safely into a land flowing with milk and honey.

When the Jewish people came into the land of Israel, the prophetic ministry of Moses was carried on by Joshua, and in the time of the Judges, the poetess Deborah served as both judge and prophet. Samuel, the most renowned and respected of the prophet-judges, initiated the monarchy by anointing Saul as the first king of Israel, and Nathan served as personal prophet to David.

Among the northern tribes of the ninth century B.C.E., Elijah is known for taking on the whole establishment in Ahab, Jezebel, and the 450 prophets of Baal, his successor, Elisha, served as prophet, political adviser, and worker of miracles.

The prophetic task in the Northern Kingdom of Israel differed from that in the Southern Kingdom of Judah. Because of its geographical location on the trade routes, the Northern Kingdom was more sophisticated, urbane, and pluralistic in its lifestyle and in its worship of the gods. The social elite of the Northern Kingdom fostered interdynastic marriages, which resulted in a plethora of gods and relied on fertility cults for prosperity in agriculture and trade. They also ignored the cries of the poor, building their social order and lifestyles on injustice and exploitation. These practices were intolerable to the prophets of the North.

Amos, one of the earliest of the classical prophets, addressed these cultic concerns and the social responsibilities of justice and concern for the poor with fierce and unsurpassed vigor:

Thus says the Lord:
For three crimes of Israel, and for four,
I will not revoke my word;
Because they sell the just man for silver,
and the poor man for a pair of sandals.

They trample the heads of the weak
  into the dust of the earth,
  and force the lowly out of the way.
Son and father go to the same prostitute,
  profaning my holy name.
Upon garments taken in pledge
  they recline beside any altar;
And the wine of those who have been fined
  they drink in the house of their god.

(Am 2:6–8)

Hosea, a contemporary of Amos, found his prophetic voice in the unrequited love of Yahweh and made his claims for covenantal fidelity and the exclusive worship of Yahweh with unsurpassed tenderness:

When Israel was a child I loved him,
  out of Egypt I called my son.
The more I called them,
  the farther they went from me,
Sacrificing to the Baals
  and burning incense to idols.
Yet it was I who taught Ephraim to walk,
  who took them in my arms;
I drew them with human cords,
  with bands of love;
I fostered them like one
  who raises an infant to his cheeks;
Yet, though I stooped to feed my child,
  they did not know I was their healer.

(Hos 11:1–4)

Hosea also perceived the relation between infidelity and injustice and the well-being of the covenant landscape:

Hear the word of the Lord, O people of Israel,
for the Lord has a grievance
against the inhabitants of the land:
There is no fidelity, no mercy,
no knowledge of God in the land.
False swearing, lying, murder, stealing and adultery!
in their lawlessness bloodshed follows bloodshed.
Therefore the land mourns,
and everything in the land languishes:
The beasts of the field,
the birds of the air,
and even the fish of the sea perish.

(Hos 4:1–3)

Influenced by the prophetic vision in the North and the eventual fall of the Northern Kingdom to Assyria, the seventh century B.C.E. prophets of the Southern Kingdom perceived the same infidelities and the same fate in the South. Isaiah, Micah, Jeremiah, Zephaniah, and Habakkuk echoed the prophecies of the North and prepared the people for disaster. Isaiah's song of the vineyard images this infidelity and destruction with haunting tenderness:

Let me now sing of my friend,
my friend's song concerning his vineyard.
My Friend had a vineyard
on a fertile hillside;
He spaded it, cleared it of stones,
and planted the choicest vines;
Within it he built a watchtower,
and hewed out a wine press.
Then he looked for the crop of grapes,
but what it yielded was wild grapes.

Now, inhabitants of Jerusalem and men of Judah,
judge between me and my vineyard:

What more was there to do for my vineyard
  that I had not done? . . .
Now I will let you know
  what I mean to do with my vineyard:
Take away its hedge, give it to grazing,
  break through its wall, let it be trampled!
Yes, I will make it a ruin:
  it shall not be pruned or hoed,
  but overgrown with thorns and briers;
I will command the clouds
  not to send rain upon it.
The vineyard of the LORD of hosts is the house of Israel,
  and the men of Judah are his cherished plant . . .

(Is 5:1–7)

These prophecies were fulfilled in 587 B.C.E., when the Babylonian armies swept into Jerusalem and reduced it to rubble. They had invaded Judah in 603 B.C.E., and again in 598 B.C.E., pillaging the land and deporting the king, but now the destruction was complete. The people who were left, those who had not fled to the hills or the caves or the neighboring lands, were gathered at a small village north of Jerusalem called Ramah and separated from one another. Some were led away to be executed, some were set free to live off the scorched land, and some were bound together in chains and led away to Babylonia.

Jeremiah, who had been left in Jerusalem, continued to support these exiles through letters, and to affirm that Babylonia was the agent of God in this disaster. He encouraged them to build houses and to plant gardens, to marry, raise children, and promote the welfare of the cities in which they lived. Ezekiel, who had been taken to Babylonia, supported the exiles in their midst. For both of these prophets, this disaster was a purification. It was not the end. Israel was still Yahweh's beloved, and there would be a remnant, a band of survivors, who would return. These poor ones, these *anawim,* would be given a new covenant and a new heart:

> The days are surely coming, says the LORD, when I will make a new covenant with the house of Israel and the house of Judah. It will not be like the covenant I made with their ancestors when I took them by the hand to bring them out of the land of Egypt. . . . But this is the covenant that I will make with the house of Israel after those days, says the Lord: I will put my law within them, and I will write it on their hearts; and I will be their God, and they shall be my people . . . they shall all know me, from the least of them to the greatest, says the Lord. . . . (Jer 31:31–34; [Ez 37:26–28] [NRSV])

In 550 B.C.E., Cyrus of Persia began a movement of military conquest across the northern territories of Babylonia. On October 13, 539 B.C.E., he entered the city of Babylon and assumed control without a struggle. The prisoners were released, and the captives were set free. Laden with provisions and gifts, they set out for their homeland.

With the end of the monarchy and the rise of a priestly rule, prophets and prophecy fell away. Without a king to be held accountable to a covenant, and with a covenant that had been broken, classical prophecy declined. With the fall of Babylonia and the rise of the Persian Empire, the apocalyptic writings came into being, and the wisdom literature began to flower.

ECO-SPIRITUAL REFLECTIONS

The Holy One of Israel

In looking to the traditions of the ancient Near East, it becomes apparent that Yahweh was not indigenous to Israel. The gods of early Israel were the El deities of El Elyon, El Shaddai, and the god of Beth-El, as well as Baal, the god of fertility, and the female deities of Ashura and the Queen of Heaven. *Yahweh* is not even a Hebrew name. It does not conform to the Hebrew *I Am,* and the Hebrew language itself did not come into being until after the tribes of Moses entered Israel. *Yahweh* is probably an Amoritic name as it conforms to the Amoritic verb root meaning *to be, I am,* and *He is.*

Yahweh came into Israel with the tribes of Moses. The Mosaic tradition remembers Yahweh as a desert God, a God of the wilderness and the nomadic people of the wilderness. It places Yahweh's home on Sinai with the Midianites and names Jethro, the Midianite father-in-law of Moses, as a priest of Yahweh (Exodus 18). Early Egyptian records support this in listing Yahweh as a god who dwells in the area of Sinai, and other evidence suggests that Yahweh was worshiped by the Kenites near Sinai from time immemorial. Through the Moses tribes, Yahweh moves from the mountain of Sinai to the mountain of Zion. Yahweh, who led the Hebrews out of Egypt, is literally carried into Israel by these tribes in the ark of covenant. Like the tribes themselves, Yahweh came into Israel as a stranger in a foreign land.

In the early days of the monarchy, the worship of Yahweh existed alongside the worship of the other gods of Israel. David, who was from the southern tribe of Judah, was a Yahwist and established Yahweh as the God of the royal cult in Jerusalem. In the northern territories of Israel, El was the prominent name of God, and the Elohist tradition was begun there. These traditions merged after the fall of the Northern Kingdom in the eighth century B.C.E., when the Elohist tradition was taken south and joined to the Yahwist tradition. In this merging, Yahweh was understood as the only God of Israel, as "The Holy One of Israel," and a monotheism began to emerge within Israel.

Monotheism, in its present context, came into being through the experiences of destruction and exile. As the prophets had foretold the fall of Jerusalem and Judah, the exile was seen as the punishment of Yahweh. It was thought that if Yahweh had sovereignty over the armies of Babylonia, then Yahweh was not only greater than all the other gods but also the universal God, involved with all the tribes and nations of the earth.[5] The prophetic writings of Isaiah and Micah envisioned a day when all nations would recognize this and stream toward the mountain of the Lord:

In the days to come,
The mountain of the LORD's house
shall be established as the highest mountain
and raised above the hills.
All nations shall stream toward it;
many peoples shall come and say:
"Come, let us climb the LORD's mountain,
to the house of the God of Jacob,
That he may instruct us in his ways,
and we may walk in his paths."
For from Zion shall go forth instruction,
and the word of the LORD from Jerusalem.
He shall judge between the nations,
and impose terms on many peoples.
They shall beat their swords into plowshares
and their spears into pruning hooks;
One nation shall not raise the sword against another
nor shall they train for war again.
(Is 2:2–4; [Mic 4:1–3])

Although Yahweh as "The Holy One of Israel" appears in Hosea, Jeremiah, Ezekiel, Habakkuk, the Psalms, and Job, it is the Isaian tradition that exalts this Mysterious Other. Beginning with the inaugural call of Isaiah of Jerusalem, this imagery appears thirty times in the Isaian writings:

In the year King Uzziah died, I saw the Lord seated on a high and lofty throne, with the train of his garment filling the temple. Seraphim were stationed above; each of them had six wings: with two they veiled their faces, with two they veiled their feet, and with two they hovered aloft.

"Holy, holy, holy is the Lord of hosts!" they cried one to the other. "All the earth is filled with his glory!" At the sound of that cry, the frame of the door shook and the house was filled with smoke [incense]. (Is 6:1–4)

For the people of Israel, to be holy was to be separate, withdrawn, or other. It was to be set apart or to participate in a numinous nonordinary reality. Arising from the experience of awe, mystery, and otherness, holiness was known in three aspects. The primal aspect was the experience of the Numinous Presence and the awe, mystery, and sense of otherness that were evoked by this encounter with the Holy. This aspect found expression in the prayer and worship of the people. The second aspect was cultic holiness, which was a response to the Numinous Presence. This aspect was honored in the times, places, people, and objects set apart as gifts or for service to the Numinous Presence. The third aspect was the moral and ethical quality made manifest by the Numinous Presence. This aspect found expression in the moral and ethical integrity of individual people and of the nation itself.

Through this understanding, the covenantal relationship required that Israel enter into and participate in the holiness of Yahweh, that it be set apart as gift and for service, and that it participate in the moral and ethical qualities of Yahweh, as they were understood by the people. This perception is expressed in the simple call to holiness from the Levitical Code:

> The LORD said to Moses, "Speak to the whole Israelite community and tell them: Be holy, for I, the LORD, your God, am holy." (Lv 19:1–2)

As "The Holy One," Yahweh was transcendent, set apart, wholly other. As "of Israel," Yahweh was immanent, near, and intimately involved with the covenanted people. Through this understanding, "The Holy One of Israel" was immanent and transcendent, intimately involved and sovereignly beyond, personal and other. This all-pervading and all-encompassing presence is celebrated most eloquently in the wisdom literature and in the Psalms:

> O Lord, you have probed me and you know me;
> you know when I sit and when I stand;
> you understand my thoughts from afar.

My journeys and my rest you scrutinize,
with all my ways you are familiar....

Where can I go from your spirit?
from your presence where can I flee?
If I go up to the heavens, you are there;
if I sink to the nether world, you are present there.
If I take the wings of the dawn,
if I settle at the farthest limits of the sea,
Even there your hand shall guide me,
and your right hand hold me fast....

Truly you have formed my inmost being;
you knit me in my mother's womb....
My soul also you knew full well;
nor was my frame unknown to you
When I was made in secret,
when I was fashioned in the depths of the earth.

(Ps 139:1–15)

An Enduring Cloud of Witnesses

Throughout human history, men and women have experienced the Holy in both its immanence and its transcendence. They have experienced a level of reality beyond the "ordinary" reality that we see with our eyes, hear with our ears, and touch with our hands. These people have come to know that beyond the world of ordinary perception lies another world, a world we often refer to as the sacred or numinous world, a world of "nonordinary" reality in which we see with the eyes of our spirits or are touched by a Numinous Presence or spiritual being. This world is not something in which we believe or for which we hope. It is, rather, a reality that we experience, that humans have always experienced, and that nonhuman creation seems to experience as well. For those who experience it, this reality is more real than our "ordinary" world, and is perceived as the matrix out of which the "ordinary" world emerges.

While people may not experience or language this nonordinary or spiritual reality in the same way, they all understand it as an encounter with the Holy and often perceive other spiritual beings—angels, demons, nature spirits, and ancestors who are present to help or hinder, to foster good or to bring forth evil. Archeological evidence indicates that even tens of thousands of years ago our human ancestors perceived this reality, burying their honored dead with artifacts and flowers and binding the bodies of some less honored dead as if they intended to prevent their return.

Most cultures have guardians or guides to assist others in journeying to this sacred world during life or at the time of death. These shamans, priests, prophetic figures, mystics, and holy women and men are ones who have already journeyed to this realm or have been caught up into or experienced incursions of the Holy in their lives. These holy ones carry the tradition forward and are honored and sought out as companions for healings, vision quests, or sacred journeys.

The people of the Judeo-Christian Scriptures attest to both this experience of the Holy and to the priestly, prophetic, and shamanic ministry of their holy men and women. The Scriptures, in fact, begin and end with these experiences. In our biblical prehistory, Adam and Eve, our mythic ancestors, walked and talked with God in the garden as an ordinary everyday experience. Noah built the ark, gathered his companion animals, endured the Flood, and participated in the re-creation of the world in his continual response to God's requests.

In early historical times, Abraham emigrated to a new land at God's request and also experienced visions in his covenant with Yahweh. His descendants Jacob and Joseph encountered God in dreams, and Joseph interpreted dreams in the court of the Egyptian pharaoh.

Moses encountered the Numinous Presence of Yahweh in the burning bush on Mount Sinai and later encountered this Holy One on the same mountain, amid lightning and storm, while receiving the Commandments for the people of Israel. Elijah also encountered Yahweh on this mountain, not in the wind or fire or quaking of the mountain, but in the still small voice when the storms had ceased. Our mythic Job discovered God in the whirlwind.

Among the classical prophets, Isaiah experienced his prophetic call in the vision of God on a high and lofty throne, surrounded by angels, and Ezekiel was called to his prophetic ministry through a similar vision as he sat among the exiles along the river Chebar in Babylonia. Amos and Micah were also called through visions, and Jeremiah and Hosea were called by the spoken Word of God.

In the Christian Scriptures, Mary and Zechariah are visited by the angel Gabriel, who announces the coming births of Jesus and John, and Joseph receives a warning to flee to Egypt and a call to return to Galilee from an angel in his dreams.

At the beginning of Jesus' ministry, we find him encountering the Holy Spirit at his baptism when the heavens opened and the Spirit descended on him. The Scriptures also tell us that angels cared for him while he was in the desert and that an angel appeared to strengthen him during his long and lonely night in Gethsemane.

After Jesus' death, Mary Magdalene encountered an angel at his empty tomb, and Saint Paul encountered the Risen Jesus on his journey to Damascus. Angels opened the prison gates for Peter and the apostles in Jerusalem and for Paul and Silas in Philippi. Stephen was gifted with a vision of the heavens as he was being stoned, and Peter saw the unclean animals lowered to him from heaven when God instructed him to receive Gentiles into the Christian community. The Scriptures themselves end with the extensive visions of John, communicated by the Alpha and the Omega, "The one who is, who was, and who is to come" (Revelation 1:8).

Reawakening to the Holy

While every age and every culture has been sensitive to this reality and has discovered the Holy within it, since the seventeenth century and the mechanical-universe theories of Descartes and Newton, the cultural consciousness of the West has considered the sense of the numinous to be illusory or even pathological, because it does not fit into rational thought or a mechanical universe. Those who experienced the Holy in nature, in dreams, in voices, or visions were dismissed as hysterics or psychotics. They learned to be silent about their experiences.

Now, however, as a new millennium draws near, our Western culture is reawakening to the presence of the Holy in our lives and beginning to speak of it again. Rupert Sheldrake reflects on this phenomenon:

> Mystical experience is usually regarded as rare, confined to a few saints, sages, and visionaries. But in fact it is surprisingly common. In surveys of random samples of the population in both Britain and the United States, over a third of the people questioned said they had been aware of "a presence or a power" at least once in their lives, and for most of them this experience was very significant.... Again and again, those who described their experience to the researchers did so with a sense of relief at being able to talk about it. For many, their spiritual or mystical experience seemed to be of supreme importance, but they were unable to discuss it with their families or friends for fear of ridicule or being thought mentally unbalanced. This research has revealed in fact that there is a widespread taboo in our society against admitting to such experience. (*The Rebirth of Nature,* pp. 213–14)

Sheldrake also reminds us of ways in which we may seek or draw near to the Holy. For Sheldrake, the first of these is through pilgrimage, through a return to our place of birth or such sites as temples, shrines, sacred trees, and holy wells that have been recognized and honored as sacred places by the community. In reflecting on pilgrimages to these places, he suggests ways in which we might enhance our experience of the Holy:

> One [way] is to become aware of the local geography and geomancy, the lie of the land, the qualities of the surroundings,

> and the life of local plants and animals. Another is to learn the stories of the place, to rediscover the local myths and to learn the names of the guardian spirits or patron saints. Yet another is to recognize the local sacred places by visiting them. And most effective of all is to open oneself in prayer to the sacred presence in the place.
>
> (*The Rebirth of Nature,* p. 219)

Sheldrake also reminds us that we can approach the Holy through sacred time, through the cycles of night and day as well as the rhythms of new moons and Sabbaths, seasonal festivals, and holy days.

While Sheldrake stresses pilgrimage as a way to approach the Holy, most religious traditions look to prayer as the primary approach. They foster the simple lifting of the mind and heart to God, in silence or in speech, in petition, gratitude, or praise.

A second way to seek the Holy that has emerged within all religious traditions is that of fasting, of emptying oneself and one's life of the nonessential to make room for the Holy. This fasting is usually understood as a partial or complete abstinence from food or even water, but it can also be restraint or abstinence in speech or various other activities.

A third way of approaching or inviting the Holy is through almsgiving, through the letting go or sharing of one's time, talents, or resources with the poor, with those who are in need, or those who cry out for mercy or compassion.

In all these ways, we draw near to the Holy and wait. And yet, through time and place, season and festival, pilgrimage and prayer, the Holy also draws near to us and waits. Ultimately, it does not matter if we find God or God finds us. All that matters is that we meet.

The Cry of the Poor

The Hebrew Scriptures have their beginnings in the cry of the poor. Enslaved and oppressed in the land of Egypt, the Hebrews cried out in anguish, and their cries reached to the heavens and the mountain of God. From the burning thorn bush of Sinai, Yahweh responded with compassion:

> I have witnessed the affliction of my people in Egypt and have heard their cry of complaint against their slave drivers, so I know well what they are suffering. Therefore, I have come down to rescue them. . . . (Ex 3:7–8)

Rooted in Israel's experience of slavery and oppression in Egypt, in its saving experience of the Exodus, and its covenantal relationship with Yahweh, this cry of the poor found its highest expression in the classical prophets of Israel and Judah in the seventh and eighth centuries B.C.E. These times before the 722 B.C.E. fall of Samaria in the North and the 587 fall of Jerusalem in the South were times of religious infidelity, social injustice, and oppression. Amos and Hosea in Israel, and Micah, Isaiah, and Jeremiah in Judah, cried out against these realities. They understood and announced, in the name of Yahweh, that these injustices would result in the collapse of the natural and social world in which they lived.

Robert and Mary Coote write of the sociopolitical climate of these times in ways that resonate with our own time in an alarming familiarity:

> The system of commercial agriculture [for export] imposed during the long reigns of Jeroboam and Uzziah had a dire effect on Palestinian villagers, especially in the highlands. Specialization of production destroyed the diversification of agriculture upon which villages in the higher lands depended for livelihood. Forced by taxation on grain into cultivation of perennial cash crops, grapes and olives, farmers could no longer practice field rotation, fallowing, planting legumes to restore nutrients in the soil, or raising livestock. Meanwhile villagers were forced to encumber their land as collateral for loans at exorbitant rates. As arable land was apparently more

> alienable than land in perennials, this was the first to fall into the hands of creditors. To feed themselves and their families, villagers then had to bring inferior marginal land into grain production, on which they got less food for more work and higher cost. Struggling with inexorable debt and taxes, small landholders lost what little land they had, and an increasing number became wage laborers and debt slaves.
>
> While the state was procuring the maximum of wheat, wine, and oil through taxes and other exactions and its own production, villagers who could not raise enough food to feed themselves were forced to borrow silver against their future harvest. . . .
>
> Meanwhile rich landowners impressed their labor to build second and third country homes and ate increasing quantities of meat. . . . The large landlords dominated local courts, where villagers might have sought relief from foreclosure. As the ever-fewer landholders became wealthier, the gap between the rich and poor widened, and social ties designed to relieve the burden of debt came apart. Absentee landowners congregated in the cities, especially Samaria and Jerusalem, and the traditional bond between patron and client deteriorated into anonymous exploitation.
>
> (*Power, Politics, and the Making of the Bible,* pp. 48–49)

The prophets' cry against these injustices articulated not only their own anguish and grief but the anguish and grief of Yahweh, the people, and the land itself. As the covenant was framed in law, the prophets often pronounced their oracles in the form of a covenant lawsuit. These oracles began with an indictment of the powerful elite who were neglecting or exploiting the marginalized and the powerless and who were insensitive to the cries of the poor. This indictment was followed by a warning or threat of the impending destruction of their natural and social world through conquest or disaster. The oracle usually concluded with a call to repentance and a return to covenant fidelity. Through repentance and return a remnant would be saved. Micah proclaims:

Listen, all you peoples,
attend, earth and everyone on it!
Yahweh intends to give evidence against you,
the Lord, from his holy temple....

You have already been told what is right
and what Yahweh wants of you.
Only this, to do what is right,
to love loyalty
and to walk humbly with your God....

Listen, tribe of assembled citizens!
Can I overlook the false measure,
that abomination, the short bushel?
Can I connive at rigged scales
and at the bag of fraudulent weights?
For the rich there are steeped in violence,
and the citizens there are habitual liars....

The faithful have vanished from the land:
there is no one honest left.
All of them are on the alert for blood,
every man hunting his brother with a net.
Their hands are adept at wrong-doing:
the official makes his demands,
the judge gives judgement for a bribe,
the man in power pronounces as he pleases....
The earth will become a desert
by reason of its inhabitants, in return for what they have done....

What god can compare with you
for pardoning guilt
and for overlooking crime?...
Once more have pity on us,
tread down our faults;
throw all our sins
to the bottom of the sea.

(Mic 1:2; 6:8–12; 7:2–3,13,18–19 [NJB])

The Contours of Compassion

Traditional biblical scholarship has understood the prophetic concern for the poor as a legal concept arising from the experience of exodus and covenant. Abraham Heschel, however, sees it as an issue of compassion, as an expression of our participation in the pathos of God. For Heschel, it is a legal aspect because it is first an issue of compassion.[6]

German theologian Gerd Theissen understands this concern for the poor as an evolution of evolution in which the biological principle of selection or survival of the fittest through competition and aggression is transcended by the cultural ethic of compassion and inclusion. This understanding honors the transformation of law from external reality to internal affair of the heart.[7]

Marcus Borg echoes this movement from legalism to compassion in terms of a transformation of consciousness. He describes these prophets as cultural critics and iconoclasts who shatter society's most cherished beliefs and give voice to an alternative consciousness:

> For the prophets, God cared about what happened in history, and not simply what happened within individuals. The concern of these God-intoxicated individuals was thus political in the sense we have used the term. That is, their concern was the *shape* of the community in which they lived. They criticized their culture's central loyalties and values and called their culture to change at a very fundamental level. As critics of culture and advocates of another way, they were concerned with their people's collective life, both in its present state and its historical direction. This concern was the intersection between Spirit and culture, God and history.
>
> (*Jesus: A New Vision,* p. 156)

These prophetic voices are still heard today in the countless men and women who respond to the cries of the poor. In our land, Dorothy Day responded to the cry of the homeless and the hungry in creating caring communities to shelter and feed them. Martin Luther King, Jr., responded to the cry of African-Americans who had been freed from slavery but not from lives that were separate and unequal. The protesting communities of Philip and Daniel Berrigan raised our awareness of the destructiveness and insanity of nuclear weapons and the arms race. César Chavez rose up amid the cry of migrant workers and brought them dignity and an improved quality of life. These and numberless others have suffered political harassment, imprisonment, and even death in their compassionate response to the cry of the poor.

The Suffering of Creation

These cries that have been heard, however, are not the only cries. There are many yet unheeded cries among the human community and among the larger community of creation. And there is the unheeded cry of our compassionate and suffering God. Contemporary theologians Arthur Peacocke[8] and Jay B. McDaniel,[9] as well as innumerable mystics throughout the ages, perceive a God who suffers in creation. They believe or know that God feels and suffers in and with the beings of the earth. Jesus affirms this, telling us that whatever we do to the least of our brothers and sisters, we do to him (Matthew 25:40).

As the young man asked Jesus "Who is my neighbor?" (Luke 10:29), we would do well to ask, "Who is my brother or sister?" Saint Francis knew the sun and moon, the wind and rain, and all creatures of the earth as his sisters and brothers. His whole Canticle of the Creatures celebrates this unity and diversity within the family of God. Native American traditions experience this same perception. Chief Sealth (Seattle), the wise prophetic elder of our land, speaks for all people who have this understanding:

> Every part of this earth is sacred to my people. Every shining pine needle, every sandy shore, every mist in the dark woods, every clearing, and humming insect is holy in the memory and experience of my people.... The perfumed flowers are

> our sisters; the deer, the horse, the great eagle, these are our brothers. The rocky crests, the juices of the meadows, the body heat of the pony, and man — all belong to the same family.... The air is precious to the red man, for all things share the same breath.... This we know. The earth does not belong to man; man belongs to the earth. This we know. All things are connected like the blood which unites one family. All things are connected. Whatever befalls the earth befalls the sons of the earth. Man does not weave the web of life, he is merely a strand in it. Whatever he does to the web, he does to himself. (Address to tribal assembly, Chief Sealth, 1854. Notes and translation by Dr. Henry Smith 1854. Taken from *Thinking Like a Mountain,* pp. 68–71)

In our day, these sisters and brothers in the community of creation are suffering deeply, and we experience their silent agony as the cry of the poor. The birds of the air are being poisoned by environmental toxins and die through loss of food resources, habitat, and migratory resting sites. The beasts of the field are being confined to enclosures, humiliated, and used as commodities. The fish of the rivers and seas perish through oil spills, the dumping of toxic wastes, brutal fishing technologies, and loss of spawning grounds. The soil of the earth is being poisoned with pesticides and fertilizers, and washes and blows away through poor agricultural practices. The forests and wetlands are being destroyed with all their beauty and biodiversity, and with all their power to contain, purify, and recycle water and carbon dioxide. The air is heavy with poisonous fumes, and tons of particulate matter rain down on our cities every day. The ozone layer thins, resulting in disease, death, and the destruction of the very basis of the planetary food chain. Our whole earth and all its beings cry out as the web of life is assaulted everywhere and strand after strand weakens and begins to fall away.

If we cannot change all these realities at once, we can, at least, begin to turn our hearts around. We can open ourselves to the sacred in our lives and in our world. We can seek wisdom and compassion. We can look to and honor our contemporary prophets and learn from the wisdom of indigenous people who have lived long and lovingly on the land. We can also foster reconciliation with these brothers and sisters in the community of creation through listening and presence, prayer and ritual, ecological sensitivity and environmental restoration. In all these ways, we turn and return, moving into a future filled with hope.

A Future Filled with Hope

One of the most significant awarenesses to come out of the experience of exile was the concept of creation as a continuing event. It arose through the experiential understanding that in spite of infidelity, failure, and collapse, God was always free to do a new thing, God was always willing to gather up the fragments and begin once more. The prophets also came to believe that out of a primal fidelity and love for creation, Yahweh would remain faithful to that which had come into being through divine wisdom and divine love.

While surrounding nations continued on within a perception of cyclical time, honoring birth and death, springtime and harvest, Israel came to understand these cyclical experiences as expressions of divine fidelity within an ongoing process of linear time, which we have come to call history. Yahweh was a God who continued to create and who could make all things new.

There is, perhaps, no better prophetic book to reflect God's continuing to create than the book of the prophet Joel. Writing about 350–400 B.C.E. in the land of Judah, Joel struggles to find meaning and hope as a plague of locusts devastates the land. The vivid language of his anguish and terror resounds across the millennia:

> Listen to this, you elders;
> everybody in the country, attend!
> Has anything like this ever happened in your day,
> or in your ancestors' day?
> Tell your children about it

and let your children tell their children,
and let their children the next generation!

What the nibbler has left, the grown locust has eaten,
what the grown locust has left, the hopper has eaten,
and what the hopper has left, the shearer has eaten....

For a nation has invaded my country,
mighty and innumerable,
with teeth like a lion's teeth,
with the fangs of a lioness....

The harvest of the fields has been lost!
The vine has withered,
the fig tree wilts away;
pomegranate, palm tree, apple tree,
every tree in the countryside is dry,
and for human beings
joy has run dry too....

The country is like a garden of Eden ahead of them
and a desert waste behind them.
Nothing escapes them.
They look like horses,
like chargers they gallop on,
with a racket like that of chariots
they spring over the mountain tops,
with a crackling like a blazing fire
they devour the stubble,
a mighty army in battle array....

At the sight of them, people are appalled
and every face grows pale.
Like fighting men they press forward....
They hurl themselves at the city,
they leap on to the walls,
swarm up the houses,
getting in through the windows
like thieves.

As they come on, the earth quakes,
the skies tremble,
sun and moon grow dark,
the stars lose their brilliance.

(Joel 1:2–12, 2:3b–10 [NJB])

Joel understands this terrifying experience of invasion and destruction as a day of judgment. The skies, darkened by the flight of locusts, truly reflect a day of darkness. The destruction of Joel's natural and social world truly is a day of doom. Joel, who was probably a priest associated with the temple, calls for a time of repentance, a time of prayer, fasting, and lamentation:

Priests, put on sackcloth and lament!
You ministers of the altar, wail! . . .
For the temple of your God has been deprived
of cereal offering and libation.
Order a fast,
proclaim a solemn assembly;
you elders, summon
everyone in the country
to the temple of Yahweh your God.
Cry out to Yahweh. . . .

(Joel 1:13–14 [NJB])

After the destruction has ended, after the repentance and return, the people, the land, and the creatures of the land enter into a time of renewal and the beginning of a new era filled with the blessings of Yahweh:

Yahweh said in answer to his people,
'Now I shall send you
wheat, wine and olive oil
until you have had enough. . . .'
Land, do not be afraid;

be glad, rejoice,
for Yahweh has done great things.
Wild animals, do not be afraid;
the desert pastures are green again,
the trees bear fruit,
vine and fig tree yield their riches.
Sons of Zion, be glad,
rejoice in Yahweh your God...

'I will make up to you for the years
devoured by the grown locust and the hopper...
You will eat to your heart's content....
And you will know that I am among you in Israel,
I, Yahweh your God, and no one else.'

(Joel 2:18–27 [NJB])

With cessation of the plague and the renewal of blessing, Joel moves from the immediacy of the experience to an understanding of the experience as a microcosm of a final or eschatological time of transformation. Using the apocalyptic imagery of his time, Joel looks to a time when the Spirit of God will be poured out on all humanity, to a planetary Pentecost and a transformed humanity that will come into being through God's wonderfully mysterious power to create and re-create out of chaos, destruction, and nothingness. For Joel, Yahweh is a God who makes all things new, and for those who have remained faithful, for those who have come to repentance, for the remnant, there will be a future filled with hope:

"After this
I shall pour out my spirit on all humanity.
Your sons and daughters shall prophesy,
your old people shall dream dreams,
and your young people see visions.
Even on the slaves, men and women,
shall I pour out my spirit in those days...."

(Joel 3:1–3 [NJB])

This book is deeply about transformation and tells us that in moving from the old into the new we go through a time of suffering and chaos, a "Day of the Lord," in which we experience a complete and total breakdown of our old natural and social world and the emergence of a new order of being. It reminds us that we move in and out of order and into eras of transformation in response to the mutually interactive freedoms of God, human creation, and nonhuman creation.

This speaks profoundly to our contemporary time in which the natural and social worlds that we have known appear to be disintegrating before our eyes, and a new world, held out to us in hope, has not yet come into being. Like Joel, we can only look forward to it, seeking and following the new thing the Lord is doing. With Joel, we can vision and dream and hope.

Chapter Four

THE WISDOM LITERATURE

Wisdom traditions are a universal aspect within the human community. They have existed throughout the world and across time in the oral traditions of tribal communities as well as in the elegant literature of highly developed civilizations. They come to us through Native Americans, Mesoamericans, the Aboriginal peoples of South America and Australia, tribal Africa, China, India, and the ancient Near East.

The wisdom tradition of Israel is one strand in this wider wisdom and finds its beginnings and its uniqueness within the traditions of the ancient Near Eastern lands of Egypt and Mesopotamia. Both historical and archeological evidence confirm Israel's dependence on the wisdom traditions of these ancient Near East neighbors.[1]

The Nature of Wisdom

Israelite wisdom arises from the lived experience of the people. It does not look to the law or the prophets, to covenant or cult, but searches for meaning and mastery in the everyday experiences of life. It looks to the natural world and the human community for laws and patterns that help people to live in harmony with one another, with the universe, and with God.

Israelite wisdom fervently believes that God is good and can be trusted. It finds evidence of divine faithfulness in the order of creation — in sunrise and sunset, in seedtime and harvest. It probes the ambiguities of conflicting truths and ponders the mystery of innocent suffering. It questions the experiences of poverty and injustice, illness and death.

Israelite wisdom is concrete and practical, and Israelite wisdom is subtle and illusive. It is not simply a body of knowledge, but a disposition of the heart. It is a perceptive sensitivity and an intuitive knowing that participates in the life of God and reveals the withinness or interiority of realities and events. Ultimately, wisdom is a gift. It is a self-revelation of the being of God, that is to be longed for and to be sought, yet when it comes, it comes as gift.

Israelite wisdom springs forth from and covers the full range of human emotions. It expresses the love and concern of a parent passing on a teaching or a skill, and the insight and puzzlement of a wise elder reflecting on the ambiguities of life. It moves from somber religious ponderings on the nature of God and the problem of suffering to exuberant expressions of gratitude, celebration, and joy in the goodness of God and the wonders of creation.

Kathleen O'Connor calls this everyday wisdom a spirituality of the marketplace and writes of its wholeness:

> Wisdom spirituality accepts, indeed, blesses, life in the marketplace. It takes everyday existence with the utmost seriousness. It asserts that ordinary human life, here and now, in all its beauty, ambiguity, and pain, are of immense importance to human beings and to their Creator. In wisdom's view the struggles and conflicts of daily life are not to be escaped but embraced in full consciousness of their revelatory and healing potential. The reason for this view is wisdom's underlying assumption that all of creation exists in the presence of its Creator. (*The Wisdom Literature,* p. 16)

The Setting of Wisdom

For Israel, as for other nations, the early core of wisdom is found in oral traditions within the family and tribe. It finds its beginnings in the teaching or instruction of a father to his son, a mother to her daughter, and parents to their children. Gathered and treasured as folk or clan wisdom, this wisdom was used to educate the young, guide the mature, and settle disputes within the larger community. The Book of Proverbs carries many of these instructions in brief, easily remembered *mashals* or teachings.

> My son, forget not my teaching,
> keep in mind my commands;
> For many days, and years of life,
> and peace, will they bring you. Let not
> kindness and fidelity leave you;
> bind them around your neck;
> Then will you win favor and good esteem
> before God and man.
> Trust in the Lord with all your heart,
> on your own intelligence rely not;
> In all your ways be mindful of him,
> and he will make straight your path.
>
> (Prov 3:1–6)

A beautiful and extended teaching within the family group is found in the story of Tobit, a story that includes prayers, psalms, and words of wisdom:

> "My son, when I die, give me a decent burial. Honor your mother, and do not abandon her as long as she lives. . . . Give alms from your possession. Do not turn your face away from any of the poor, and God's face will not be turned away from you. . . . Do not keep with you overnight the wages of any man who works for you, but pay him immediately. . . . Give

to the hungry some of your bread, and to the naked some of your clothing.... Seek counsel from every wise man and do not think lightly of any advice that can be useful.... "

(Tobit 4:3–19)

A later setting for the teaching of wisdom in Israel is the royal court. Throughout the ancient Near East, rulers assembled wisdom scholars and scribes to assist them in diplomacy, in ruling wisely, and in recording the activities and wise sayings of the court. Israel was no exception to this practice. David gathered a variety of counselors for his court, and Solomon is renowned for his employment of professional sages from Egypt and Babylonia. It was these sages and scribes who gathered, edited, or created the early wisdom literature of Israel.

A third setting for the wisdom tradition may have been schools of wisdom both within and apart from the court. In the court, sages and scribes tutored the children of the king and may have also conducted schools for the children of the wealthy within the court milieu. There is also evidence that professional wise men conducted classes in speech, wisdom, diplomacy, and money-lending for the children of the elite outside the court.

Within the traditions of Israel, Solomon is known as the wisest of all. We are told that he prayed for and received wisdom, and that he was praised for his wisdom. The writings also honor him for his discernment, his gathering of great scholars, his wise diplomacy, his great knowledge of the natural world, and his wealth. In reflecting on the life of Solomon, however, it becomes evident that while he is held up as the consummate wise man and a number of wisdom writings are attributed to him, his life did not always display a great deal of wisdom.

In reality, Solomon lived and died hundreds of years before most of the writings attributed to him were written. This was an acceptable practice in the ancient Near East. Ascribing authorship to a renowned person was a way of situating a writing within a particular tradition and establishing its authority. Another reason for his reputation as a wise man is that in the early proverbs wisdom was rewarded by wealth. Because Solomon was the wealthiest man of his time, he was also considered to be the wisest. A third reason for this discrepancy between the man and his reputation is that the history of Solomon as the exemplar wise man was enhanced or perhaps written back into the Scriptures when they were reworked in the court of Hezekiah. Another reason may be our human need to project our unfulfilled ideals and dreams and our unrecognized potentials and possibilities on someone or something beyond us. In early Israel, because a king was viewed as a corporate personality who embodied the whole of Israel, this projection would have come more easily. Therefore, Solomon is undoubtedly considered wise, not only because of the wisdom he did have, but because he carried the projected ideals, dreams, and yearnings of the people of Israel, even across time.

The Many Faces of Wisdom

Though expressions of wisdom can be found throughout the Hebrew Scriptures, some books belong entirely to the wisdom tradition. The earliest of these are Proverbs, Job, and Ecclesiastes. Written in Hebrew and completed after the time of exile, these books are found in both the Catholic and Protestant Scriptures. The later books of Ecclesiasticus and the Wisdom of Solomon were written in Greek during the Hellenistic period. These later books belong to a collection known as the Apocrypha and are found only in the Catholic and Eastern Orthodox Scriptures.

An assortment of other writings are closely related to these books and are accepted as wisdom literature. The Song of Songs, attributed to Solomon, finds its place through Solomon's name as well as through the everyday experience of joy and delight in human love. Several psalms (1, 23, 34, 37, 49, 73, 111, 112, and 128) are also accepted as wisdom literature, and many stories of the wise, such as the stories of Joseph, Esther, and Daniel, are also included and held up as models in this literature.

The themes of these wisdom writings are as multifaceted as life itself and move through time in an evolutionary way. In their simplest beginnings, they focus on the practical aspects of living wisely, on doing good and avoiding evil, on acquiring wealth and receiving blessing, and on living harmoniously in the human community and in the cosmos:

> Better a poor man who walks in his integrity
> than he who is crooked in his ways and rich.
> Without knowledge even zeal is not good;
> and he who acts hastily, blunders. . . .
> He who keeps the precept keeps his life,
> but the despiser of the word will die.
> He who has compassion on the poor lends to the Lord,
> and he will repay him for his good deed.
>
> (Prov 19:1–3, 16–17)

The later wisdom builds upon, questions, and reinterprets the earlier traditions. Confronted with ambiguity, injustice, and evil, it moves into deep and poetic ponderings on the mysteries of life, on innocent suffering and evil in the presence of a God who is good, and on the ultimate destiny of the just in a life beyond the one we know:

> But the souls of the just are in the hand of God,
> and no torment shall touch them.
> They seemed, in the view of the foolish, to be dead;
> and their passing away was thought an affliction
> and their going forth from us, utter destruction.
> But they are at peace.
> For if before men, indeed, they be punished,
> yet is their hope full of immortality;
> Chastised a little, they shall be greatly blessed,
> because God tried them
> and found them worthy of himself.

As gold in the furnace, he proved them,
  and as sacrificial offerings he took them to himself.
In the time of their visitation they shall shine,
  and shall dart about as sparks through stubble...
Those who trust in him shall understand truth
  and the faithful shall abide with him in love:
Because grace and mercy are with his holy ones,
  and his care is with his elect.

(Wis 3:1–9)

In its most sublime moments, the later wisdom enters into the perceptive and intuitive edges of the mystery of God through the personification of Wisdom, the subtle and elusive one who came forth from God, who was with God in the beginning, who delights in the human ones, and who abides in the heights, the depths, and the withinness of creation:

The Lord created me at the beginning of his work,
  the first of his acts of long ago.
Ages ago I was set up,
  at the first, before the beginning of the earth.
When there were no depths I was brought forth,
  when there were no springs abounding with water.
Before the mountains had been shaped,
  before the hills, I was brought forth—
when he had not yet made earth and fields,
  or the world's first bits of soil.
When he established the heavens, I was there,
  when he drew a circle on the face of the deep,
when he made firm the skies above,
  when he established the fountains of the deep,
when he assigned to the sea its limit,
  so that the waters might not transgress his command,
when he marked out the foundations of the earth,
  then I was beside him, like a master worker;

and I was daily his delight,
  rejoicing before him always,
rejoicing in his inhabited world
  and delighting in the human race.

(Prov 8:22–31 [NRSV])

At every level, these themes are relevant for our time.

ECO-SPIRITUAL REFLECTIONS

The Mystery of God and the Creation of the World

The most profound reflections within the wisdom literature center on the mystery of God, the magnificence of creation, and God's continuing presence in creation. This mystery of God is expressed as ultimate mystery in Job's search for meaning, when God answers with nonanswers, inviting Job to let go and surrender to the Mysterious One and the mystery.

Then the LORD addressed Job out of the storm and said . . .
Where were you when I founded the earth?
  Tell me, if you have understanding.
Who determined its size; do you know?
  Who stretched out the measuring line for it?
Into what were its pedestals sunk,
  and who laid the cornerstone,
While the morning stars sang in chorus
  and all the sons of God shouted for joy? . . .

Have you ever in your lifetime commanded the morning
  and shown the dawn its place . . .
Have you entered into the sources of the sea,
  or walked about in the depths of the abyss? . . .
Have you fitted a curb to the Pleiades,
  or loosened the bonds of Orion? . . .

Do you hunt the prey for the lioness...
Do you know about the birth of the mountain goats...
Do you give the horse his strength...
Does the eagle fly up at your command...

(Job 38–39)

This search for God, for our origins, and for meaning amid mystery has continued to occupy philosophers, theologians, astrologers, scientists, and sages throughout human history. In our own day, the scientific community of the late twentieth century has given us a new cosmology and a new understanding of how things came into being. This scientific understanding does not invalidate the mythic or biblical understanding of creation in any way. Rather, it enhances our understanding and appreciation and magnifies our wonder, awe, and praise for the creativity and wisdom of the Mysterious One who has brought all of creation into being.

A New Story of the Universe

The scientific story tells us that our universe had its origins in the primal flaring forth of the fireball.[2] It tells us that the universe emerged as an infinitesimal point of pure energy, a primal particle of infinite density and no dimension, that expanded in a micromoment to a small symmetrical ball — a cosmic egg, perhaps the size of a hazelnut, that any one of us could hold in the palm of our hand. The cosmic egg then flared forth, foaming out in fire, and the universe was born.

In our day, we have begun to see back to the primordial fireball, to its very edges, to the light that was there 16 billion years ago and is observable today as dim background glow at microwave frequencies. But even this does not answer all our questions about the mysterious beginnings of the universe.

As far as we know, the universe emerged out of apparent nothingness, in the same way that elemental particles emerge from apparent nothingness in our present theory of quantum physics.

It flamed into being out of mystery, out of a realm beyond our understanding, from the tendencies and allurements in a field of divine consciousness, as a self-expression of Ultimate Mystery.

Though we cannot see back beyond the primal flaring forth because time and space form the perimeters of our existence, scientists continue to speculate about the origins of the fireball itself. Many scientists believe the universe came into being through fluctuations in a background vacuum, from a field or matrix. Other scientists think the universe may have emerged from an opening or a white hole from another universe, much as the sand in an hourglass slips through from one sphere to another. Still others speculate that this cosmic-energy event occurred when the universe emerged in its four dimensions from the fracturing of an unstable ten-dimensional universe. Many physicists believe this tremendous energy event was enhanced by the mutual annihilation of primitive particles of matter and antimatter. Still others theorize that this primordial energy event was caused by fluctuations within the primal point of energy.

Whatever its origins, in the first fiery moments of its existence the fireball began to cool and the first elemental particles began to emerge — quarks, leptons, photons — primal energy freezing into a thick, opaque cosmic soup. As the fireball continued to cool, protons and neutrons formed from the gathering of quarks, and primitive nuclei formed and were torn apart, again and again, in the intense heat and activity. Then, three minutes after the beginning, stable nuclei formed and remained.

The fireball burned for perhaps a million years, expanding outward, not into anything, but simply expanding, finite and unbounded, creating time, creating space. As it cooled, slowly, slowly, the expanding light began to stretch from minute gamma rays to rays of infrared, to a wavelength that ushered in what we have come to call the epoch of darkness — though there were yet no eyes to know the darkness or the light.

During this time of cooling, energy continued to freeze into particles of matter, and particles came together as atoms of hydrogen and helium. These atoms gathered in gravitational allurement, streams of gossamer clouds, clustering, falling in on themselves, warming, circling, brightening for 3 or 4 billion years until the giant quasars flamed into being and light returned to the heavens.

With the passing of time, the great galaxies emerged, the simple ellipticals and the exquisite spirals, which birth stars and star clusters in the density waves of their spiral arms. All these fill the heavens by the billions, moving outward, in ever-growing loneliness, in ever-growing communion, still creating time, still creating space.

Our own earth came into being in the arms of the great galaxy we have named the Milky Way. Born with our solar system from the stellar debris of a great star in its violent dying, Earth has evolved as a living being and has blossomed forth all the life we know to exist in the whole of the universe. For those who have seen her from space, whose lives have been forever altered by her radiance, she is a blue-white jewel with clay-red land masses, set in the deep black silence of space. And she is our home.

The Mysterious One and the Mystery

While the contemporary scientific community looks to the empirical dynamics of the universe, religious communities have always looked to the mysteries of the universe and the Mysterious One who brought it into being. In the cosmology of the biblical tradition, the biblical writers portray God creating the universe out of chaos. This imagery arises in the time of exile and reflects the cosmologies and Creation stories of the ancient Near East. The concept of creation as continuing emerges about the same time within the prophetic tradition. Creation out of nothing, which has been the dominant position within the Christian tradition, first appears in the book of Maccabees and does not really gain a hearing until the patristic period and the work of Augustine.

For Augustine, all things came into being through God's action. There was no primal matter or chaos before the beginning. There was only God. Augustine also understood creation as continuing in his concept of "seed-principles" in which all things were present in the beginning in potential, awaiting a favorable time of appearance.[3] This concept is quite consistent with the epigenetic unfolding of the new cosmology and contemporary evolutionary theory. It differs mainly in its nonrecognition of the interplay of necessity and chance.

At the present time, systematic theologians Philip Hefner, Jürgen Moltmann, and Wolfhart Pannenberg support a theology that affirms creation out of nothing and continuing creation.[4] They affirm the sovereignty of God and the absolute dependence of all creation on God as well as continuing creation through the indwelling Spirit. They also find significant consonance with the scientific cosmology of the universe emerging from a primal singularity in the cosmic fireball.

Process theologians, such as Charles Birch, John B. Cobb, Jr., and David Griffin, affirm the biblical tradition and support creation out of chaos and continuing creation.[5] They hypothesize God creating out of the chaos of the scattered remnants of the fireball as well as God creating out of his or her own inner chaos. The God of process thought images a God who is neither all-powerful or powerless. God is, rather, participating in creation. Often presented in feminine imagery, he or she is present in the unfolding or birthing of creation as well as in every existential moment, not intervening but offering possibilities. In exploring this process thought, Ian Barbour writes:

> The God of process thought is neither omnipotent nor powerless. Creation occurs throughout time and in the midst of other entities. God does not predetermine or control the world, but participates in it at all levels to orchestrate the spontaneity of all beings... God does not intervene sporadically from outside, but rather is present in the unfolding of every event. Creative potentialities are actualized by each being in the world, in response both to God and to other beings. The process view emphasizes divine immanence, but it by no means leaves out transcendence.
>
> ("Creation and Cosmology," *Creation as Cosmos*, p. 144)

A third voice arises in the writings of paleontologist Teilhard de Chardin and geologian Thomas Berry, who perceive a psychic or spiritual dimension within the whole of creation. This reflects the intuitive perception of indigenous and tribal peoples as well as mystics of all traditions throughout the ages.

Teilhard experiences the Christian story as identical to the universe story. He believes that the universe had, from its beginning, both a physical-material dimension and a psycho-spiritual dimension. Teilhard experienced Christ in the heart of all matter, and perceived all creation moving together toward a final fullness in the Omega point of the Cosmic Christ.

Berry perceives the universe story as our sacred story and the universe as our sacred community. He believes the theological community needs to realize this if it is to contribute to the healing of our contemporary culture. Berry also perceives three governing principles within the universe that express its inner life and reflect the persons and activities within the sacred community of God.[6]

The first of these principles is differentiation, a reality that began in the primordial fireball with the formation of the first elemental particles and the separation of the four universal forces. It has continued on in the emergence of quasars and galaxies, suns and moons, mountains and mists, butterflies and fireflies. It finds expression in the reality that no two beings are identical and that every being is unique, from galaxies to snowflakes to human beings. Every era fosters this multiplicity and diversity of individual beings and realities within the universe.

The second principle in Berry's cosmology is subjectivity. This principle affirms the primal psychic dimension, the interiority or withinness, in all beings and realities that has increased with the unfolding of the universe and the greater complexification of being. This subjective and psychic dimension has also found acceptance in the scientific community through the work of Werner Heisenberg. Heisenberg, discovered that the observer influences the behavior of seemingly inanimate matter, a reality that suggests an inner freedom within the heart of all matter.

Berry's third principle is communion. This principle is understood as an interdependence and mutual indwelling in which every being and reality in the universe is coextensive and in communion with every other being and reality. He perceives the universe as a single multiform and celebratory energy event, where everything is connected to everything else and all things are bonded together in love.

Berry understands this trinity of principles as a reflection of the Sacred Community of God. He sees the principle of differentiation as a reflection of the Father, the subjectivity or inner articulation of creation as a reflection of the Son, and the intercommunion or bonding in love as the reflection of the Holy Spirit.[7]

A fourth perspective, one beyond creation out of chaos, creation out of nothing or continuing creation, is that of God as lure from the future. It finds a place in the work of Wolfhart Pannenberg and Ted Peters. Peters believes that the future has an ontological priority over the past or the present, a priority that is inherent in the origin and destiny of the universe and provides a source of freedom to become:

> We experience anxiety regarding what is to come, wondering whether or not we will be here to share the future or even if there will be a future for anybody. The power of being is preconsciously apprehended as that which can overcome anxiety by assuring the future. To be is to have a future. To lose one's future and to have only a past is to die; and deep down we know it. The dynamic perdurance of the present moment is contingent on the power of the future to draw us into it. The first thing God did for the cosmos was to give it a future. Without a future it would be nothing.
>
> ("Cosmos as Creation," *Cosmos as Creation,* p. 87)

Pannenberg, in reflecting on the Stoic doctrine of pneuma and contemporary field theory within the scientific community, perceives God as the field from which the universe emerges:

> The turn toward the field concept in the development of modern physics has theological significance. This is... because field theories from Faraday to Einstein claim a priority for the whole over the parts. This is of theological significance, because God has to be conceived as creator and redeemer of the world. The field concept could be used in theology to make the effective presence of God in every single phenomenon intelligible.
>
> ("The Doctrine of Creation and Modern Science," *Cosmos as Creation,* p. 164)

This theory of God as primal field understands the Sacred Community of God to be present in the original moment of creation, in the always and everywhere of continuing creation, and beyond the edges of the final fullness of creation. In this way, God is understood as the Alpha and the Omega, the One who is, who was, and who is to come.

The Feminine Face of God[8]

A beautiful and eloquent woman emerges in the writings of the Wisdom Literature and grows in wisdom and grace as these Scriptures unfold. Beginning in the Book of Proverbs as the Wise Woman who summarizes the early wisdom of Israel, this beautiful and poetic figure grows in power and presence through time, and rises from the pages, not only to become real, but to become an exquisite Living Being who carries within herself the very spirit, reality, and life of God.

Coming into these writings after the Exile as an addition to the early wisdom literature, this Wisdom Woman[9] is contrasted with the Foolish Woman who leads men astray and eventually to their death. In these earlier writings, she appears among the people, calling out in the streets, and inviting all people to the banquet of abundance and life. Subtle and elusive, her words and manners echo the words and manners of the prophetic voice and call the human ones to live in the divine pattern of wisdom that is inherent in creation. Like the unknown prophet of the Exile, she proclaims oracles of salvation, promising life and blessing and relationship to those who seek her, love her, and follow in her ways:

> I am calling you, all people,
> my words are addressed to all humanity.
> Simpletons, learn how to behave,
> fools, come to your senses....
>
> I love those who love me;
> whoever searches eagerly for me finds me.
> With me are riches and honor,
> lasting wealth and saving justice.
> The fruit I give is better than gold, even the finest,
> the return I make is better than pure silver.
> I walk in the way of uprightness
> in the path of justice,
> to endow my friends with my wealth
> and to fill their treasuries.
>
> (Prov 8:4–21 [NJB])

The Wisdom Woman also appears as a mysterious spiritual being who dwells with God and who is deeply involved with creation. Singing a song of herself, she proclaims:

> I came forth from the mouth of the Most High,
> and covered the earth like a mist.
> I dwelt in the highest heavens,
> and my throne was in a pillar of cloud.
> Alone, I compassed the vault of heaven
> and traversed the depths of the abyss.
> Over the waves of the sea, and over all the earth,
> and over every people and nation I have held sway....
> Before the ages, in the beginning, he created me,
> and for all the ages I shall not cease to be.
>
> (Sir 24:3–9 [NRSV])

Subtle, elusive, and poetic, her whole being participates in ambiguity. In her elusiveness, she is Sovereign Presence. In her solitary calling out, she is Ultimate Relationship. In the ecstasy of her poetic lyricism, she permeates all things. Like the Noachic rainbow, she is the exquisite and elusive bridge between heaven and earth, between God and the human, and through her presence as the withinness of all things, she holds all beings and realities together in the sacred community of creation:

> For in her is a spirit
> intelligent, holy, unique,
> Manifold, subtle, agile,
> clear, unstained, certain...
> Firm, secure, tranquil,
> all-powerful, all-seeing,
> And pervading all spirits,
> though they be intelligent, pure and very subtle.
> For Wisdom is mobile beyond all motion,
> and she penetrates and pervades all things by reason of her purity.
> For she is an aura of the might of God
> and a pure effusion of the glory of the Almighty...
> For she is the refulgence of eternal light,
> the spotless mirror of the power of God,
> the image of his goodness....
> And passing into holy souls from age to age
> she produces friends of God and prophets....
> Compared to light she takes precedence...
>
> (Wis 7:22–30)

Wisdom scholar Kathleen O'Connor understands this relational aspect as her primary mode of being:

> The primary mode of being of the Wisdom Woman is relational. In all the texts where she appears, the most important

> aspect of her existence is her relationships. Her connections extend to every part of reality. She is closely joined to the created world; she is an intimate friend of God; she delights in the company of human beings. No aspect of reality is closed off from her. She exists in it as if it were a tapestry of connected threads, patterned into an intricate whole of which she is the center. (*The Wisdom Literature,* p. 59)

O'Connor continues with the blessings that come through following this alluring woman:

> To follow the Wisdom Woman, to become wise, is to awaken to, and to participate in, this matrix of relationships. It is to take a communal, holistic stance toward the world and its inhabitants, to live in communion with all that is. It is to leave behind the illusion of isolation, that we live alone, that our personal safety is all. To follow the Wisdom Woman is to make a choice for life in its most complete and wholesome possibility. (*The Wisdom Literature,* p. 60)

The historical and literary origins of the Wisdom Woman are uncertain. O'Connor believes the Wisdom Woman came into being as the male projection of the perfect woman. Many feminists, such as Joan Chamberlain Engelsman, see in her a strong resemblance to the goddesses of Israel's neighbors, such as Inanna of Sumeria, Isis of Egypt, and Demeter and Persephone of Greece. Eleanor Rae and Bernice Marie Daly understand the Wisdom Woman as the feminine aspect of the divine that is revealed in both Word and Spirit. Susan Cady understands the Wisdom Woman as a response to Israel's changing sociocultural milieu, introduced through the spread of Hellenistic culture:

> It was the drive to keep things connected that was at the heart of the wisdom tradition. In the face of threats to Israel's national consciousness and to its provincial view of the world,

> the wisdom tradition sought to create a new, more connected frame of reference. While groups within the priestly tradition in Israel and Judaism sought to separate and reisolate the Hebrew faith, the wisdom tradition was trying to integrate the Hebrew perspective into the larger picture. It is probably not accidental that the figure who imaged this effort to connect and integrate was Sophia [Wisdom] herself. Although Sophia did not represent in any major way the reemergence of women in Hellenistic society, she was the primary symbol for the human connecting enterprise, an intrinsic component of feminist spirituality as we have understood it. Because Sophia symbolizes connectedness on a number of levels, she provides a promising starting point for the development of a powerful mythic figure at the heart of feminist spirituality.
>
> (*Wisdom's Feast,* pp. 54–55)

Feminist Spirituality and the Way of Wisdom

Whatever her origins, the Wisdom Woman finds a profound resonance with contemporary feminist spirituality in several areas of concern and offers an exquisite mythic image not only for women's spirituality but for *human* spirituality and cultural transformation.

As the wisdom literature itself looks to everyday experience as the source of both ethical and philosophical wisdom, the Wisdom Woman articulates her self-understanding and claims her authority through her own experience. She has been present and active in the creation of the cosmos. She has experienced its heights and its depths. She dwells with God, and she makes her home with humanity. She is withinness and presence and motion. She delights in creativity and play.

In a similar way, within feminist spirituality the inner self is understood as the source of authority, and wisdom is acquired through sensitive listening and reflection on one's own experience as an embodied being. Like the Wisdom Woman herself, the woman of feminist spirituality grows through the moment-by-moment transformation of self and continually moves toward radical authenticity and a willed congruence with Ultimate Reality.

A second area of resonance is that of interconnectedness or relationship. As the wisdom tradition looks to both the wholeness of creation and the uniqueness of individual beings and realities, the Wisdom Woman celebrates the uniqueness and interrelatedness of all things. As the one who permeates all things and holds all beings and realities together, she is deeply embedded in the web of relationships as the matrix of these relationships, and celebrates this interconnectedness in the union of heaven and earth, heights and depths, the human and the divine.

In a similar way, feminist spirituality celebrates the interconnectedness of all creation, and the unity and diversity within this web of relationships. Like the Wisdom Woman, it cries out its desire for all human beings and all creation to participate in the banquet of abundance and life — women and men, people of color, the young and the elderly, plants and animals, forests and streams, water and soil and atmosphere. Feminist spirituality seeks a renewed perception of creation as our sacred community, and a renewed connection of humanity with the earth from which we have emerged. Feminist spirituality fosters the equality and the well-being of all, and cultivates consensus and compassion. It seeks to integrate immanence and transcendence, nature and culture, reflection and action, knowledge and mystery, perceiving them not as dualities to be overcome, but as dialectics that give life, energy, motion, and wholeness.

A third area of resonance is that of sharing power. In the wisdom tradition, power is acquired through wisdom and living wisely, through moving with the divine order that is inherent in creation. The Wisdom Woman embodies and shares this power freely, giving wisdom, life, and blessing to all who love her, seek her, and follow in her ways. Active in the creation of the world, and permeating and moving within all things, she is herself the wisdom and divine ordering inherent in creation.

Feminist spirituality also fosters this sharing of power. Through its understanding of interconnectedness and relationship, it fosters pluralism and community and invites all persons to articulate their own experiences and share their unique gifts and talents within the web of relationship. It honors the powers of listening, waiting, and visioning, and calls each person to come out of the shadows of oppression and embody creativity, authenticity, wisdom, justice, compassion, and peacemaking.

These ways of being in the world are antithetical to Western patriarchal consciousness, which experiences authority as an external reality to be feared, obeyed, or desired, and seeks transcendence and disembodiment through reason, knowledge, and law. This patriarchal consciousness is not to be understood as "men's spirituality" in either a past or contemporary sense. It is, rather, an interlocking system of psychological, sociological, and spiritual myths, symbols, and values that have been validated in societies dominated by men and internalized by men and women alike. Based on a hierarchical model, it places supreme value on men at the expense of women, children, the animal kingdom, and, ultimately, the earth itself. It encourages heroic singularity through competition, conquest, and physical, psychological, and spiritual violence, and gains power over through domination, control, oppression, coercion, and exploitation. It seeks to divide and conquer, to fragment and manage, reducing wisdom to knowledge and knowledge to information. In its fear of intimacy and its quest for power, it establishes faceless nonrelational social forms that are separate and unequal.

In an age when our human societies are being ruptured through alienation and anomie, and our biotic communities are being decimated through environmental insensitivity and greed, the feminist values of inner authenticity, interconnectedness, and the sharing of power are vital to our survival as a people and as a planet. In cultivating them, as individual persons and as a people, we will surely find ourselves moving closer to a human and a planetary wholeness, and an era of justice, compassion, and lasting peace.

Living Wisely

In our beginnings, we humans looked to the heavens in wonder and awe, to the earth in astonishment and praise. In our dawning consciousness, we somehow knew, however dimly, that we were here to acknowledge the beauty and terror, the loneliness and communion, of the universe and the Mysterious One who brought it into being. We also knew, though not so dimly, that everything was alive with mystery and everything was worthy of awe, for all our world was mystical and magical, and everything was holy.

In our primitive societies, the Divine Presence was everywhere, in mountains and rivers, animals and trees, as well as in the natural phenomenon of wind, rain, fire, and storm. This all-pervading Numinous Presence gave strength and order and meaning to our lives and enabled us to gather around this Numinous Presence as our sacred center, living in a mystical awareness of the unity of all being and moving with the seasons and the tides, the natural and spiritual rhythms of the heavens and the earth.

We have moved out of this primitive sensing and ordering through the great classical civilizations into an age of technological dominance and spiritual impoverishment. We have moved to a distant place, far, far from our sacred center where we live over and against our natural world and its processes, envisioning earth and nature apart from God and apart from the human, substituting science and technology for myth and mysticism, creating ecological and psychic disruptions, setting loose our destructive powers on the earth and the beings of the earth.

Through this alienating movement, we have entered into a time of crisis that not only affects the human but places the whole earth community in peril. This crisis is greater than any we have ever known. Our waters are filled with toxic wastes and our air with deadly pollutants. Our soil is being depleted, and our forests are dwindling. It is not the natural world that has betrayed us, however. It is we who have betrayed the natural world. Our environmental crisis is, first and foremost, a crisis of the human spirit, and of the will to live simply and peaceably as a creature among other creatures on the earth.

Geologian Thomas Berry writes that these issues are not simply about survival but involve a mode of being and a quality of spiritual presence:

> We have before us the question not simply of physical survival, but survival in a human mode of being, survival and development of intelligent, affectionate, imaginative persons thoroughly enjoying the universe about us, living in profound communion with one another and with some significant capacities to express ourselves in literature and creative arts. It is a question of interior richness within our own personalities, our shared understanding with others, and of a concern that reaches out to all living and non living beings of earth, and in some manner out to the distant stars in the heavens.
> (*The Dream of the Earth,* p. 37)

Berry understands our present way of inhabiting the earth as an adolescent mode and envisions an ecological age in which we will come into a mature wisdom and embrace our unique and profound responsibility for the earth:

> We are involved in a process akin to initiation processes which have been known and practiced from earliest times. The human community is passing from its stage of childhood into its adult stage of life. We must assume adult responsibilities. As the maternal bonds are broken on one level to be reestablished on another, so the human community is being separated from the [its] dominance of Nature on one level to establish a new and more mature relationship. . . . Now the earth insists that we accept greater responsibility, a responsibility commensurate with the greater knowledge communicated to us. (*The Dream of the Earth,* pp. 47–48)

This time of transition or transformation is both a time of turmoil and a moment of grace. It is a time of breakdown and breakthrough, and amid our reasons for despair, we have reason to have hope. As we near the end of this century and the beginning of a new millennium, we see a radical newness emerging all around us. The military is beginning to beat its swords into plowshares, disarming its nuclear weapons, reducing its personnel, and turning its resources and concern to the protection and well-being of the oppressed and the afflicted. Economists are beginning to reevaluate the meaning of wealth, moving away from a myopic economy based on competition and profit to one that includes quality of life and social well-being in terms of literacy, health, education, environmental quality, biodiversity, cultural diversity, income distribution, job satisfaction, family, and friends. Economists are even coming to the realization that indicators of wealth, such as the gross national product, need to be replaced by indicators that include both internal and external costs and the burden laid upon the whole earth community in terms of resource depletion, environmental degradation, ecosystem destruction, and waste management.

Attentive to the signs of the times, politicians are beginning to lay aside partisan interests to form a new, nonviolent political will that will serve all people and create radical systemic changes in environmental ethics, population issues, economics, and world order. Through the new physics and the rediscovery of mysticism, scientists are beginning to discover anew the spiritual dimensions of reality, and theologians are turning their eyes from an unknown future in a heavenly realm to a radical appreciation of the natural world. Everywhere we turn we are looking for sustainability and even beyond, to a future-oriented process environmentalism that is alive with the spiritual dimensions of reality.

In moving into this new eco-spiritual age, we need to remember our sense of the sacred, our original integrity and wholeness. We need to repent of our insensitivity and violence to the earth and the beings of the earth. We need to return to our natural and spiritual wisdom, gathering once again around our sacred center, sharing once more in the sacred web of life.

In living wisely, we need to look around at the destruction we have inflicted on our world, in our lakes and streams, our seas and soil and atmosphere. We need to look at our care-less-ness in the toxic wastes that engulf us, in the species that disappear through our mindlessness, in the urban scarring we create, in our waste of human lives and human potential. We need to remember our original love, the blue-green jewel of the earth, the splendor of the night skies, the warm company of our brothers and sisters, the magnificence of the Mysterious One who has loved us into being.

We need to repent of our insensitivity and violence, to confront and own and embrace our personal, national, and collective human shadow, acknowledging our darkness and the darkness we have created around us. We need to enter into the pain and mourning of the earth and the species of the earth, into the dark night of the earth that we have brought about in this time and place, in this corner of the universe.

We need to return to an original relationship with the earth, to make a covenant of peace with the earth and the creatures of the earth, with the water and soil and atmosphere, with the species we have held in no esteem, with our brothers and sisters of other cultures and creeds, with those who have gone before us, and those yet to be born, to whom we leave a wounded and diminished earth.

We need to return to the wisdom of the earth and the earth process, taking our lessons, once again, from the birds of the air and the flowers of the field, from the seasons and the tides, learning wisdom and patience and living in harmony with the earth and the beings of the earth.

We need to bring the wisdom of diverse tribes and peoples, both past and present, into all our dreams and realities as part of the collective wisdom of earth consciousness. We need to accept responsibility to be the listening heart and the compassionate consciousness of the planet, moving always toward an ever-greater consciousness, an ever-deepening response of love.

We will not be alone in this. We have never been alone. We are inexorably bound to our compassionate and faithful Creator whose love does not come to an end, who still calls us to relationship and responsibility in an evocative cry of passion, tenderness, and hope.

Part Two

THE CHRISTIAN SCRIPTURES

Chapter Five

THE GOSPELS

At the heart of the Judeo-Christian Scriptures lie four writings of incomparable power, and at the heart of these writings we discover, again and again, the "compelling beauty of Jesus" of Nazareth.[1] While we usually think of these writings as four gospels, there is, in reality, only one gospel, one magnificent announcement that comes to us through the unique perspective of four different authors or evangelists. This gospel or "Good News" announces that in Jesus of Nazareth there has been a radical inbreaking of the reign of God, and that a new and transformed reality has come into being.

These gospel writings are neither histories nor biographies. They are, rather, portraits of Jesus, sketched by their authors for their unique communities and the needs of these unique communities.[2] Donald Senior writes:

> The Gospel as written by Mark or Matthew or Luke or John was not composed for the wide world or to be included in some universal collection of church literature. The Gospel writers had a much more immediate purpose. They gathered together the traditions about Jesus available to them in their particular locale and put them together into a coherent story, a literary whole, in such a way that it would speak eloquently to the problems and hopes of the community of Christians they served. (*Jesus: A Gospel Portrait,* p. 22)

Senior also tells us that these evangelists had a role similar to that of the prophets. Having a strong sense of the past, they too sought to apply their early Christian heritage to situations within their contemporary community:

> As men of faith, they drew on the tradition of the church and shaped it in such a way that it spoke boldly and eloquently to the present. The tradition they used was not theological speculation about the meaning of Christian life or musings about the nature of God or Jesus but traditions about Jesus' life, his words, his actions, the reactions of disciples, opponents, and crowds. These were put together in such a way that each Gospel portrait of Jesus had a deep impact on the community for whom it was written. (*Jesus: A Gospel Portrait,* p. 22)

These writings emerged over time through a process that responded to the evolving experience of the Christian community. They had their beginnings in Jesus' own announcement that the reign of God was at hand, and in the excitement of the people as the "Good News" of Jesus' teachings, healings, and miracles spread through towns and villages and across the countryside of Galilee.

They developed in a more formal way as an oral tradition after the death of Jesus with the Good News of the Resurrection and the central theological proclamation that Jesus was the Son of God, the anointed one, who had come to initiate the reign of God and who would return in glory. Over time, the stories of Jesus, his teachings, healings, and miracles, were added to this central theological proclamation, and this Good News was proclaimed to both Jew and Gentile.

The final stage, the actual writing of the gospels, is exceedingly complex and cannot be traced with certainty. It is thought that the earliest protogospel was a collection of the teachings, healings, exorcisms, and miracles of Jesus that is known as *Quelle,* Source, or simply Q. It is also thought that the gospels according to Mark and John were written in their earliest form at this time, and that all of these early writings came into existence without any dependence on one another.

In the years that followed, the gospels according to Matthew and Luke were written in their earliest form, with both Matthew and Luke building on Mark and *Quelle* as well as sources that were uniquely theirs. Later, as these writings became more widely known and underwent revisions, they became interactive and interdependent. Matthew, Mark, and Luke, in fact, are so similar when placed side by side that they are known as the synoptic gospels, or those that are similar "at a glance." The Gospel of John is quite different from these gospels, and arises from its own independent source. Biblical scholars, however, believe its final redactor was familiar with these other writings and may have drawn upon them in some ways.

Mark

The Gospel according to Mark is generally thought to be the first gospel to be completed. Though its author is unknown, it has been traditionally accepted as the writing of John Mark, a companion of Peter, who gathered the teaching and preaching of Peter in Rome. As it is written in Greek and explains Jewish customs and Aramaic phrases, it is assumed to be written for a community of Gentile Christians. Because it mentions the 70 C.E. destruction of the temple in Jerusalem, it is thought to have come into its final form just after 70 C.E.

The author of the Gospel of Mark centers on the humanity of Jesus, on the "Son of Man" who is also the "Son of God," and on Jesus as the anointed one, the hidden and unrecognized Messiah who must suffer and die. Perhaps addressing a community under persecution, Mark portrays a Jesus who is misunderstood and finds his destiny through the cross and the Resurrection.

Philip Van Linden finds three important themes within the gospel, themes that are applicable to the Christian community in any age. The first of these is the humanity of Jesus. The second is trust as the heart of discipleship. The third is service to others, even as Jesus was the suffering servant of all. Van Linden writes:

> Of the four Gospel portraits of Jesus, Mark's is by far the one that best reveals the *human side of Jesus.*... Mark's readers will sense that the Jesus of this Gospel is very approachable, because he has experienced life as they have, with all its disappointments and its loves, with all its joy and sadness.... Mark believes that the truest sign of being Jesus' disciple is *trust.*... Although Mark's Gospel does not give long lists of "how to" serve God and others, its readers cannot avoid the model of Mark's Jesus as the suffering servant of all. They know that they must seize every opportunity to serve others in charity if they want to be his followers.
>
> (*The Gospel According to Mark,* pp. 8–9)

Matthew

The Gospel according to Matthew is thought to have arisen within the community of the apostle Matthew and to have come into its present form around 85–90 C.E. Completed by a second generation Jewish Christian in Antioch of Syria, for a community that was predominantly Jewish, it centers on the reign of God and is rich in images from the Hebrew Scriptures. It stresses the continuity between the Israel of the Hebrew Scriptures and the Christian community that has gathered around the risen Jesus. It portrays Jesus as the fulfillment of these Scriptures, particularly the law and the prophets, and continually expresses the relationship between the traditions of Israel and the newness of Jesus.

The structure of Matthew's Gospel is magnificent. Divided into seven major areas, it begins with the birth and infancy of Jesus and closes with the death and empty tomb of Jesus. Between these events lie five major "speeches" of Jesus. Constructed out of the traditional teachings of Jesus, Matthew arranges these teachings according to themes: the Sermon on the Mount (Ch. 5–7), the Missionary Discourse (Ch. 10), the Parables Discourse (Ch. 13), the Discourse on the Christian Community (Ch 18), and the Eschatological Discourse (Ch. 24–25).

The principal theme of the gospel is the reign of God. It announces that Jesus is the New Moses, the Son of God, the Son of David, and the Messiah. It also proclaims that in Jesus, God has drawn near to us, and the final fullness of God's reign has begun. This gospel also looks to the Christian community as the family of God, and upholds Peter as the foundational leader of the Christian community. It calls the community to humility, mutual love, and limitless forgiveness.

Luke/Acts

The Gospel of Luke is the first of a two-part work that begins with the infancy of Jesus and follows the emerging Christian community as it is led by the Spirit into the heart of the Gentile community in Rome. The Book of Acts, in fact, is so closely aligned to the Lucan Gospel and so filled with the activity of the Spirit that it is also known as the Gospel of the Holy Spirit. Both books resound with the theme of "journey," led by the Spirit.

Though both the identity of the author and the date of composition are uncertain, most scholars continue to affirm the author as Luke, a companion of Paul in his later journeys and during his imprisonment in Rome. Almost all scholars affirm the author as a Greek-speaking Christian who completed this work in Antioch or Asia Minor between 80–90 C.E.

Luke's writing is thought to be addressed to Gentile Christians scattered throughout the Roman Empire. These men and women are predominantly middle and upper-class urbanites who are unfamiliar with Palestine and the Jewish tradition. Luke writes to affirm their inclusion in the saving work of Jesus and writes his Acts of the Apostles as a semihistorical gospel of inclusion.

The innumerable themes of the Gospel of Luke are filled with grace, from the infancy narratives announcing the births of John and Jesus to the blessing of the disciples as Jesus ascends into the heavens. One of these themes is universal salvation. Drawing on the Hebrew tradition, Luke proclaims life and blessing to all nations through the ascension of Jesus and the releasing of the Spirit into the church and into the world. He expresses his understanding that the Christian faith flows forth from Judaism and out to the farthest reaches of the Roman world as an expression of God's will that all peoples enter into the new community of God.

A second theme is compassion. This compassion is a grace and mercy that is extended to all people — to sinners, outcasts, women, and the poor. It finds expression in the stories of the Loving Woman (7:36–50), the Good Samaritan (10:29–37), the Prodigal Son (15:11–32), the Pharisee and the Tax Collector (18:9–14), the Lost Sheep (15:3–7), and Jesus and the Children (18:15–17). This gentle grace, made flesh in Jesus, is held out as a pattern to be followed by all who would embrace the community of Jesus.

A third theme is joy. This joy arises through the Spirit and finds expression in prayer and praise. It flows through the gospel from the canticles of Mary and Zechariah (1:46–55; 1:67–79) to the praise of the disciples awaiting the Spirit after Jesus' ascension into the heavens (24:50–53). Embracing this theme, the gospel begins and ends with the experience of "joy" in the temple.

The Acts of the Apostles continues this salvation history through the ministry of Peter and Paul and the spread of Christianity from Jerusalem to the world of the Gentiles in Asia Minor and Rome. Without this second book, we would have little information about the early church. William Kurtz writes:

> Acts is the only book in the New Testament which continues the story of Jesus into the early church. If it were not for Acts, we would have only isolated pieces of information about the beginnings of the church. We would have to dig these bits and pieces out of the New Testament letters, but would have no framework into which to put them. Acts has provided a framework for understanding not only the infor-

mation it contains but facts gleaned from Paul's letters and other New Testament books. (*The Acts of The Apostles,* p. 5)

John

The fourth canonical gospel, the Gospel according to John, has its foundations in the experience of the Beloved Disciple, John. Like the earlier gospels, this writing evolved over time and is thought to have come into its final form after 90 C.E. through a gifted theologian and writer in the Johannine community of Ephesus. A thoroughly Jewish gospel, familiar with rabbinic thought and Jewish religious practice, it builds the ministry of Jesus around the feasts and festivals of the Jewish liturgical year and illumines the liturgical and sacramental life of the early church. Unlike the synoptic gospels, which reveal the words of the historical Jesus, the Gospel of John reveals the words of the Risen Jesus, who speaks through his Spirit who has been released into the world.

This gospel begins with an elegant prologue that portrays Jesus as both Wisdom and the Word, and introduces the many themes and theological motifs that flow through the gospel. Coming into being, perhaps as a hymn within the community of John, it sings of a Beloved and Eternal One who has entered into human form to bring humanity a profound knowledge of God:

In the beginning was the Word:
the Word was with God
and the Word was God.
He was with God in the beginning.
Through him all things came into being,
not one thing came into being except through him.
What has come into being in him was life,
life that was the light of all people;
the light shines in the darkness,
and darkness could not overpower it. . . .

He was in the world
that had come into being through him,
and the world did not recognize him.
He came unto his own
and his own people did not accept him.
But to those who did accept him
he gave power to become children of God,
to those who believed in his name
who were born not from human stock
or human desire
or human will
but from God himself.
The Word became flesh,
he lived among us,
and we saw his glory,
the glory that he has from the Father as only Son of the Father,
full of grace and truth. . . . No one has ever seen God;
it is the only Son, who is close to the Father's heart,
who has made him known.

(Jn 1:1–18 [NJB])

The main body of writing within this Gospel is divided into two books, the "Book of Signs" and the "Book of Glory." In the "Book of Signs," John writes of Jesus' ministry in terms of "signs" rather than miracles. These "signs" reveal the deeper meaning of events and are followed by discourses that invite the hearer to enter more deeply into the mystery of Jesus and his oneness with the Eternal God. They also tell us that the final fullness is already present in Jesus, not as an external event, but as an internal reality that can be experienced by the believer.

In the "Book of Glory," John writes of the passing over of Jesus and the coming of the Spirit. These writings include the beautiful imagery of Jesus washing the feet of his disciples and the lyrical discourses of the Last Supper. They conclude with the death and resurrection of Jesus in which John understands the crucifixion of Jesus as his lifting up and exaltation into "glory." Throughout both books, John portrays Jesus in rich symbolic imagery reflecting the Wisdom discourses of the Hebrew Scriptures. For John, Jesus is Bread and Vine, Shepherd and Gate, Way and Truth, Life and Light.

Though an epilogue (Ch. 21) was added after the gospel was completed, perhaps to affirm the devotion and primacy of Peter and the gospel's rootedness in the Beloved Disciple, the original closing explicates the meaning and purpose of the gospel:

> Now Jesus did many other signs in the presence of [his] disciples that are not written in this book. But these are written that you may [come to] believe that Jesus is the Messiah, the Son of God, and that through this belief you may have life in his name. (Jn 20:30–31)

ECO-SPIRITUAL REFLECTIONS

A Writing on the Heart

In looking to the gospel portraits of Jesus, we realize that we know little of Jesus' appearance. We have no knowledge of the contours of his face, the strength of his gaze, or the sound of his laughter. We know nothing of the way he moved or the gestures he made. We know, however, the disposition of his heart. We know his fidelity to truth, his hunger and thirst for justice, and his compassion for the poor. From these deep remembrances of his uncommon integrity, from the quality of his living and the courage of his dying, the gospel writers came to understand him as the initiator of a new covenant and the fulfillment of the law, the prophets, and the wisdom tradition. They also came to understand that this fulfillment and radical newness did not come through a scrupulous obedience to the law or the prophets or the wisdom tradition, but through an interiority that revealed a writing on his heart.

Jesus and His Time

Jesus, as we know, grew up in Galilee, in the hilly green northern part of Palestine that engaged in agriculture. According to the gospels, he made one or perhaps three journeys to Jerusalem in Judea.[3] This southern land, as in ancient times, was dry, barren, and predominantly pastoral. Between Galilee and Judea lay Samaria, the mountainous land of the despised Samaritans who had been excluded from worship at the Jerusalem temple.[4]

Palestine, in the time of Jesus, was a small country dominated by a Hellenistic culture and a Roman rule. Its life, however, was similar to its earlier times. In the north, most people lived in small towns and villages and cultivated fields surrounding their villages. In the south, herdsmen lived in the open, tending their flocks and moving about to find pasture. There was a still a clear distinction between Jew and Gentile, and the righteous were vastly superior to the unclean. The oldest son continued to have priority over younger sons, and men were valued more than women.

Though Jerusalem and its temple continued on as the religious and cultural center of the nation, the local synagogue, which had come into existence during the time of exile, was the ordinary setting for religious gatherings. The Torah remained the embodiment of the law, and the prophets were duly honored with words and monuments. People also clung to their ancient wisdom, believing that the way of the righteous would lead to wealth and long life and the way of folly would lead to ruin and death. Some people believed in an afterlife, and some believed that this life was all they had.

Four major religious groups were prominent in the Jewish community of this time. The Sadducees were a conservative party who believed that what had been written in the Torah was binding for all time, and that one led a religious life by continuing to read and follow it scrupulously. They did not believe in spiritual beings or an afterlife. The Pharisees were a more progressive group. While they have gained a bad reputation through the gospels, they were a devout and faithful people, and both Jesus and Paul came from this tradition. The Pharisees believed that the law needed a continual interpretation within its cultural milieu and according to the signs of the time. They also believed in angels and in a life after death.

A third group or movement was that of the Zealots. They looked to the coming of a political messiah and the overthrow of Roman rule. The Zealots offered religious resistance and engaged in sporadic guerrilla warfare. The fourth major group was the Essenes, who had withdrawn to the desert lands near the Dead Sea to lead a life of religious purity. They lived a simple, communal, monastic life that centered on the Scriptures and purification rituals.

All of these groups were in conflict with the world around them. Their social and cultural world clashed with the Hellenistic culture, and their economic life was devastated by the taxation of Rome. They lived in a time of continuing crisis that seemed to have no resolution and no way out.[5] Jesus, however, came into an alternative consciousness. Grasped by the Spirit of God, he looked into his own human heart and found a realm that was not of these worlds.

Jesus and the New Consciousness

The history of Israel is filled with charismatic prophets and leaders who encountered God and lived their lives in the presence of God. Abraham heard the call of God in Ur, and followed God throughout his life. Jacob encountered God in his angelic dream of the great ladder between heaven and earth, and Moses, the great prophet and lawgiver, came into a direct experience of God on the mountain of Sinai through the burning bush. The great classical prophets Isaiah and Ezekiel encountered God in visions of angels and a heavenly throne.[6]

Jesus followed in this charismatic tradition of Israel and came into a profound encounter with God in his baptism by John. Matthew, Mark, and Luke all write of this encounter:

> It happened in those days that Jesus came from Nazareth of Galilee and was baptized in the Jordan by John. On coming out of the water he saw the heavens being torn open and the Spirit, like a dove, descending upon him. And a voice came from the heavens, "You are my beloved Son; with you I am well pleased." (Mk 1:9–11 [Mt 3:13–17; Lk 3:21–22])

After this profound religious experience, Jesus was led into the desert by the Spirit to reflect on and struggle with the call he experienced. This reflection and struggle continued throughout his life.

In searching through the gospels, it seems that Jesus' earliest self-understanding was that he was a prophet. According to Luke, when Jesus returned from the desert, he entered the synagogue and identified himself as being in the prophetic tradition through the Isaian passage announcing a ministry to the afflicted and a time of jubilee and renewal:

> He came to Nazareth, where he had grown up, and went according to his custom into the synagogue on the sabbath day. He stood up to read and was handed a scroll of the prophet Isaiah. He unrolled the scroll and found the passage where it was written:
>
> "The Spirit of the Lord is upon me,
> Because he has anointed me
> to bring glad tidings to the poor.
> He has sent me to proclaim liberty to captives
> and recovery of sight to the blind,
> to let the oppressed go free,
> and to proclaim a year acceptable to the Lord."
>
> Rolling up the scroll, he handed it back to the attendant and sat down, and the eyes of all in the synagogue looked intently at him. He said to them, "Today this Scripture passage is fulfilled in your hearing." (Luke 4:16–21)

This initial understanding as prophet is echoed in the gospels of Matthew and Mark where Jesus begins his ministry with the prophetic call to "Repent, for the Reign of God is at hand!" (Matthew 4:17; Mark 1:15). For Jesus, this reign was the inbreaking of God in both the inner world of the human heart and the outer world of the present social and political order. Like the Jubilee year, it would be a time of renewal when God would restore all things to their original integrity.

As his ministry deepened and grew, Jesus' self-understanding also deepened and grew. He came to understand that, as a prophet, he was like the prophet of the Isaian tradition, a servant who would suffer and die because of his love for God and his fidelity to his unique vision. Again, Matthew, Mark, and Luke affirm him as both the Suffering Servant and the Messiah of the Isaian tradition:

> Now Jesus and his disciples set out for the villages of Caesarea Philippi. Along the way he asked his disciples, "Who do people say that I am?" They said in reply, "John the Baptist, others Elijah, still others one of the prophets." And he asked them, "But who do you say that I am?" Peter said to him in reply, "You are the Messiah." Then he warned them not to tell anyone about him. [After this] he began to teach them that the Son of Man must suffer greatly and be rejected by the elders, the chief priests and the scribes and be killed and rise after three days.
>
> (Mk 8:27–31 [Mt 16:13–21; Lk 9:18–22])

In understanding himself as a prophet and in coming from the rabbinic tradition of the Pharisees, Jesus also understood himself as an interpreter of the law, as one who stood within the tradition of the law and yet transformed it with a radical newness. This understanding, which flows through the gospels, finds its most comprehensive expression in Matthew's eloquent Sermon on the Mount:

> Do not think that I have come to abolish the law or the prophets. I have come not to abolish but to fulfill. Amen, I say to you, until heaven and earth pass away, not the smallest letter or the smallest part of a letter will pass from the law . . . I tell you, unless your righteousness surpasses that of the scribes and the Pharisees, you will not enter into the kingdom of heaven. (Mt 5:17–20)

For the Jewish community, the greatest law was love, and had always been love. There was no change to this teaching over time. What constituted love, however, was another matter. For the Jewish community, love consisted of being holy as God was holy. It required what Marcus Borg calls a "politics of holiness,"[7] a separation from the unclean elements and beings in the world around them and a scrupulous fidelity to the detailed prescriptions of the law.

Jesus stood against this politics of holiness and called for a "politics of compassion."[8] He associated with tax collectors and prostitutes, beggars and lepers, the lame and the blind. He spoke of God's love for the good and the bad, the just and the unjust. He warned the Pharisees that their politics of holiness lacked justice, compassion, and inclusion and would lead to disaster as it had in the days of their ancestors.

Perhaps the strongest expression of Jesus' outrage against this legalistic holiness is found in the Gospel of Matthew:

> Woe to you, scribes and Pharisees, you hypocrites. You pay tithes of mint and dill and cumin, and have neglected the weightier things of the law: judgement and mercy and fidelity. These you should have done, without neglecting the others. Blind guides, who strain out the gnat and swallow the camel.
> (Mt 23:23–24)

For Jesus, the law was rooted in compassion. It was not an outward display of righteousness, but an inner disposition of the heart that flowed outward in loving-kindness and mercy, to the poor, the lame, the blind, and the sinner.

The importance of this inwardness and the "habits of the heart" also flows through Jesus' ministry as a teacher of wisdom. Like the earlier sages of Israel, Jesus taught in proverbs that were based on the everyday experiences of life, and drew his lessons from the natural world. Unlike the earlier sages or the sages of his own time, who taught from the authority of the tradition, Jesus spoke out of his own authority, out of his own experience, perception, and authenticity. His teachings are filled with stories and parables that lead from ordinary experiences into the deep wisdom of God. His reflections on the natural world reveal an unparalleled transparency to nature, and an unparalleled call to trust in the ultimate graciousness and compassion of God:

> Therefore I tell you, do not worry about your life, what you will eat [or drink], or about your body what you will wear. Is not life more than food and the body more than clothing? Look at the birds in the sky, they do not sow or reap, they gather nothing into barns, yet your heavenly Father feeds them. Are not you more important than they? . . . Learn from the way the wild flowers grow. They do not work or spin. But I tell you that even Solomon in all his splendor was not clothed like one of them. If God so clothes the grass of the field, which grows today and is thrown into the oven tomorrow, will he not much more provide for you, O you of little faith. (Mt 6:25–34)

In all his teachings, as prophet, sage, and interpreter of the law, Jesus looked to the human heart. He knew it was the inwardness that mattered. In the Jewish tradition, the heart was the deepest center of being, the source of feeling, perception, intention, and activity. It was the heart that Jesus sought and addressed, that he called to repentance and renewal. Jesus also came to understand that it was through the heart that one entered into a deep and transformed relationship with God, and into the new covenant promised by Jeremiah, the covenant that was written on the heart:

> The days are surely coming, says the LORD, when I will make a new covenant with the house of Israel and the house of Judah. It will not be like the covenant that I made with their ancestors when I took them by the hand to bring them out of the land of Egypt — a covenant that they broke, though I was their husband, says the LORD. But this is the covenant that I will make with the house of Israel after those days, says the LORD. I will put my law within them, and I will write it on their hearts; and I will be their God, and they shall be my people. No longer shall they teach one another, or say to each other, "Know the LORD," for they shall all know me, from the least of them to the greatest, says the LORD; for I will forgive their iniquity, and remember their sin no more.
>
> (Jer 31:31–34 [NRSV])

Jesus understood this transformed heart as the deepest expression of God's own being, as the only heart that could bear the imprint of God's own heart. It was a heart that could act justly, love tenderly, and walk humbly with its God. It was a heart that could be faithful as God was faithful, compassionate as God was compassionate, and full of God's own joy.

Cultivating the Transformed Heart

In the book of Micah, the prophet announces the way of the transformed heart:

> You have already been told what is right
> and what Yahweh wants of you.
> Only this, to act justly,
> to love tenderly
> and to walk humbly with your God.
>
> (Mic 6:8 [translation unknown])

The evangelists came to understand Jesus as the bearer of this transformed heart, as the initiator of a new covenant, and as the fulfillment of the law, the prophets, and the wisdom tradition. If we take some liberty with this, we might discover that to fulfill the law, one needs to pursue justice or to act justly. To fulfill the prophets, one needs to live a life of compassion or to love tenderly. To be filled with wisdom, one needs to live simply and peaceably in the community of creation or to walk humbly with one's God.

Expanding this holy trinity of justice, compassion, and inclusion to the community of creation, we also find a resonance with Thomas Berry's trinity of principles that govern the universe. We find that just as Berry's principles of differentiation, subjectivity, and communion are not confined to the human community, or even the earth community, but to the whole of the universe, the principles of justice, compassion, and inclusion are not confined to the human community, or even the earth community, but may be understood as governing principles that are inherent in creation. Because of this, through our fidelity and conscious living out of these writings on the heart, we move with the evolutionary unfolding of the universe and God's continuing intention to create, moving always and everywhere into an ever-greater congruence with Ultimate Reality.

To act justly, then, is to honor the differentiation inherent in creation, and graciously celebrate the richness of this diversity as it is expressed in the pluriformity of race, color, gender, ethnicity, cultural diversity, biodiversity, and geodiversity. It is to offer a sensitive openness and a profound concern for justice to the human and nonhuman world, to biotic communities and nonliving forms. To act justly is to listen attentively to the cries for justice from humans and nonhumans, mountains and hills, rivers and streams, fruit trees and vines. It is to recognize that, though we live in a world where beings draw life from one another, where plants live on the sun and animals live on plants and other animals, there is still a moral imperative to obtain, protect, and sustain the quality of life for all human and nonhuman beings in both their physical-material needs and in their psycho-spiritual needs. This respect for differentiation and diversity not only honors the Creator and the creature, but also enhances joy within the community of creation and leads each being into its deepest fulfillment.

To love tenderly, in this perception, is to honor the subjectivity within all beings and realities, to respect their psychic unity and their capacity for interiority. It is to experience kinship and compassion, and to feel with them in both their joy and in their grief. To love tenderly is to offer a contemplative presence to creation and to be open to the manifestation of the Holy in each being and reality. It is to call each being by a name that is truly its own, and to know each being as a "thou."

Berry writes of our need to foster this deep awareness of the sacred within each reality of the universe:

> There is an awe and reverence due to the stars in the heavens, the sun, and all heavenly bodies; to the seas and the continents; to all living forms of trees and flowers; to the myriad expressions of life in the sea; to the animals of the forests and the birds of the air. To wantonly destroy a living species is to silence forever a divine voice. Our primary need for the various lifeforms of the planet is a psychic, rather that a physical, need. The ecological age seeks to establish and maintain this subjective identity, this authenticity at the heart of every being.... Only such a comprehensive vision can produce the commitment required to stop the world of exploitation, of manipulation, of violence so intense that it threatens to destroy not only the human city, but also the planet itself.
>
> (*The Dream of the Earth*, pp. 46–47)

Finally, to walk humbly with one's God is to freely and joyously enter into the existential communion of these beings of the universe, from the vast galaxies of the heavens to the most minute creatures of the earth. It is to know that everything is connected to everything else, and to abide in this unity with magnanimity and trust. To walk humbly with one's God is to experience one's identity through mutually enhancing relationships and to find one's own place in this community with gratitude and joy. It is to live in this community in voluntary simplicity, attentive to the quality of life in the human community, in biotic communities, and in the community of earth. Deep ecologist Bill Devall writes of the richness of this simple lifestyle:

> Voluntary simplicity is not self-denial but a more compassionate approach to living and consideration for the vital needs of other creatures. Simple living does not mean involuntary deprivation, enforced austerity, boring or tedious daily routines, or poverty of experiences. On the contrary, voluntary simplicity is often a necessary condition for maximum richness, intensity, and deepness of experience.... Quality of life, individually or collectively, is measured by psychological well-being and, from a deep ecology perspective, by the well-being of the whole biotic community.
>
> (*Deep Ecology,* pp. 84–85)

These writings on the heart impel us to follow in the way of Jesus and cultivate the transformed heart. They call us to embrace the whole of creation, in every aspect of its being, with justice, compassion, and a deep sense of community. They call us to embrace the heavens and the earth, the darkness and the light, suffering and joy, life and death and even the enemy, all with equanimity and trust.

Embracing the Enemy

Among the many teachings of Jesus that we find in the gospels, perhaps the most radical ones are the call to unlimited forgiveness and the need to extend love, even to the enemy. In the Sermon on the Mount, Jesus looks to the ancient law of retaliation and says:

> You have heard that it was said, "An eye for an eye and a tooth for a tooth." But I say to you, offer no resistance to one who is evil. When someone strikes you on [your] right

> cheek, turn the other one to him as well. If anyone wants to go to law with you over your tunic, hand him your cloak as well. Should anyone press you into service for one mile, go with him for two miles. . . .
>
> (Mt 5:38–41 [Ex 21:22–25; Dt 19:21; Lv 24:19–20])

Addressing the Essene teaching to love all the "sons of light" but hate all the "sons of darkness," Jesus develops the teaching on retaliation still further:

> You have heard that it was said, 'You shall love your neighbor and hate your enemy.' But I say to you, love your enemies, and pray for those who persecute you, that you may be children of your heavenly Father, for he makes his sun rise on the bad and the good, and causes rain to fall on the just and the unjust. . . . So be perfect [compassionate] as your heavenly Father is perfect [compassionate].
>
> (Mt 5:43–48 [Lk 6:27–36])

In this gospel, Jesus also tells us that we are to forgive, not seven times, but seventy times seven (Mt 18:21–22), that we are to attend to the beam in our own eye before we attempt to remove the speck in the eye of another (Mt 7:4–5), and that if we have offended anyone, our call to reconciliation is to take precedence over our call to worship (Mt 5:23–24).

In our own lives, this call to love and limitless forgiveness usually begins with the acceptance of our own weaknesses and failings, and those of the people who are close to us and whom we love. When we live in communities, in marriages, and in families, we hurt one another again and again, usually without intending to do so. We don't always listen well. We have unreasonable expectations. We make unreasonable demands.

Relationships are messy. If we learn to live together at all, with any depth or authenticity, we do it through forgiveness — through mutual love, honesty, openness, respect, trust, and forgiveness.

Ultimately, forgiveness sustains, deepens, and authenticates community.

But there are also times when we are outside of community, when there is no mutuality possible, when we need to forgive, if only from our side, and if only for our own sake. Perhaps we do this because no real relationship ever existed, or the other is no longer available, or is an institution or a military power.

We often find ourselves forgiving one way, for our own sake, the abusive parent who scarred our lives, the drunken driver who killed our child, the academic superior who stole our work, the named or unnamed offenders in poverty and war.

Virgil Elizondo, writing in *Concilium* says:

> The greatest damage of an offense — often greater than the offense itself — is that it destroys my freedom to be me, for I find myself involuntarily dominated by inner rage and resentment. I do not recognize my own self [and] I hate the offender . . . but in the very hatred of the other I allow them to be master of my life. The ultimate sinfulness of the sin itself and its greatest tragedy is that it converts the victim into the sinner. . . . The scars made to the heart eat away at the life of the victim. Alone we do not seem to be able to rehabilitate ourselves.
>
> ("I Forgive But I Do Not Forget," *Concilium: Forgiveness,* 1986, pp. 69–79)

In the gospels, we read that the Jews were scandalized when Jesus forgave sins, because they believed that only God could forgive sin. There is a deep wisdom in this, though it may need to be understood in another way. If only one forgives, if the other is not open to or available for reconciliation, the one who forgives must absorb the pain, the injury, and the diminishment of the offense. He or she must carry and transform the anger, the resentment, and the rage. This is not possible for the ego-self. The ego-self is not strong enough to hold, absorb, and transform the injury or the injustice. It cries out for vengeance, for an eye or a tooth. It lashes out and tears itself on the memory of the wound, or buries the pain and anger in the body and carries it in illness. Only the transformed heart, the heart that carries the writing, that abides with the Transforming Spirit, is great and strong and compassionate and free enough to absorb the pain, dissolve the rage, and transform the injustice.

Across the centuries, this limitless forgiveness and this profound love of one's enemies has been addressed by a multitude of women and men who have carried within themselves this transformed heart. In our own twentieth century, one of the greatest of these was Mahatma Gandhi. Born in 1869 amid the Buddhist, Jain, and Hindu religions of India, Gandhi came into an early and deep understanding of *ahimsa* (noninjury to sentient beings) through his experience of being forgiven by his father for a small theft. This understanding deepened during his time of study in England when he came into a heartfelt resonance with the Sermon on the Mount and its teaching on nonresistance and nonretaliation. After he completed his studies, he began to live out this nonviolent resistance as a young lawyer of color in apartheid South Africa.

Through these experiences, Gandhi came to embrace and embody the Jain belief in *ahimsa* or harmlessness, the Buddhist value of universal compassion, and the Christian ethic of nonviolence and love of one's enemies. To Gandhi, this way of being was not simply a personal spirituality, but a force for social and political transformation. He also believed the power and ability of his nonviolence was directed and sustained by God (Rama), to whom he had given his life in love and service:

> If one has pride and egoism, there is no non-violence. Non-violence is impossible without humility. My own experience is that whenever I have acted non-violently I have been led to it and sustained in it by the higher promptings of an unseen power. Through my own will I should have miserably failed.
>
> (*Non-Violence in Peace and War,* Vol. I, p. 187. Quoted in Thomas Merton's *Gandhi on Non-Violence,* p. 36)

Near the end of his life, Gandhi came to believe that nonviolence had failed in India because it had never become an authentic disposition of the heart for his followers. In writing of Gandhi's life, Thomas Merton addresses Gandhi's later understanding:

> In Gandhi's mind, non-violence was not simply a political tactic which was supremely useful and efficacious in liberating his people from foreign rule, in order that India might then concentrate on realizing its own national identity. On the contrary, the spirit of non-violence sprang from *an inner realization of spiritual unity in himself.* The whole Gandhian concept of non-violent action and *satyagraha* is incomprehensible if it is thought to be a means of achieving unity rather than as *the fruit of inner unity already achieved.*
>
> Indeed this is the explanation for Gandhi's apparent failure (which became evident to him at the end of his own life). He saw that his followers had not reached the inner unity that he had realized in himself, and that their *satyagraha* was to a great extent a pretense, since they believed it to be a means to achieve unity and freedom, while he saw that it must necessarily be the *fruit of inner freedom.*
>
> (*Gandhi on Non-Violence,* p. 6)

Within a few brief years after Gandhi's death, this inner freedom was not only achieved but perfected in a later follower of Gandhi. Born as the son of a Baptist minister and moving into the ministry himself, Martin Luther King, Jr., built his whole life around the insights of Jesus and the strategies of Gandhi. In his early life, King had been moved by the great call to love and to nonviolent resistance in the teachings of Jesus. He had understood this call, however, as an interaction between individual persons. Through Gandhi, King was awakened to nonviolence and love as effective forces for social reform.

In his speeches and writings, King continually addressed six characteristics of nonviolent resistance.[9] The first of these is that nonviolence is not passive or cowardly. It is an active resistance to evil, and it resists because it is strong and has overcome fear. A second is that nonviolence does not seek to defeat or humiliate the opponent but rather seeks to initiate understanding and friendship. A third characteristic is that the resistance is not directed against the person but against the evil. The struggle is not between people but between justice and injustice.

A fourth characteristic of nonviolent resistance is its willingness to suffer and even die without retaliation. It is willing to accept violence but will never inflict it. It understands suffering as redemptive, perceiving it as a powerful force for education and transformation. A fifth reality is that it avoids not only the external violence of injury, but also the internal violence of hate. Nonviolence is always governed by love of the other and works to fulfill the other's need for human authenticity. The sixth characteristic of nonviolent resistance is the understanding that the universe is on the side of justice and that the Creative Force of the universe seeks to bring every aspect of the universe into a community of justice and peace.

These themes of nonviolent resistance and love flow eloquently through Dr. King's writings and speeches. Steeped in the biblical traditions, his life and work embody his vision of a universe filled with justice, compassion, and peace, and his enduring belief that love is the most durable power:

> Always be sure that you struggle with Christian methods and Christian weapons. Never succumb to the temptation to becoming bitter. As you press on for justice, be sure to move with dignity and discipline, using only the weapon of love. . . .
>
> In your struggle for justice, let your oppressor know that you are not attempting to defeat or humiliate him, or even pay him back for the injustices that he has heaped upon you.

Let him know that you are merely seeking justice for him as well as for your self. . . .

Honesty impels me to admit that such a stand will require willingness to suffer and sacrifice. So don't despair if you are condemned and persecuted for righteousness sake. Whenever you take a stand for truth and justice, you are liable to scorn. . . .

I still believe that love is the most durable power in the world. . . . This principle stands at the center of the cosmos. As John says, "God is love." He who loves is a participant in the being of God.

("The Most Durable Power," *A Testament of Hope*, pp. 10–11)

Chapter Six

THE NEW TESTAMENT LETTERS

The New Testament Letters are a diverse collection of writings that preserve the early Christian community's understanding of Jesus and the meaning of his life for those who came to believe in him. Written in the early church during the last half of the first century, these letters contain developing theologies and developing Christologies. They reflect on who Jesus is in himself, who he is in relation to God, how he fulfills the Hebrew Scriptures, and the nature of his place in the cosmos. These letters, which are sometimes addressed to specific persons or communities and sometimes addressed to the entire church, also address a multitude of moral and pastoral concerns within the early Christian community.

One of the greatest of these concerns in the early church, both theologically and pastorally, was the second coming of Jesus. Just as there had been great messianic expectations before the coming of Jesus, there were great apocalyptic expectations in both the Jewish and Christian communities in the first century. Among the Christians, this took on special significance after the death of Jesus.

In the early Christian community, there was a prevailing belief that Jesus would return soon and bring the reign of God into its final fullness. As Jesus had not returned, there was a continuing struggle to understand this delay and to explain it within the community. Because of this, many Christians were uncertain about how to live in this time of waiting and looked to the Apostles for wisdom. Paul addresses this issue with encouragement and hope in what is probably the earliest letter of the New Testament, the First Letter to the Thessalonians:

> Now concerning the times and the seasons, brothers and sisters, you do not need to have anything written to you. For you yourselves know very well that the day of the Lord will come like a thief in the night. . . . So then let us not fall asleep as others do, but let us keep awake and be sober . . . and put on the breastplate of faith and love and for a helmet the hope of salvation. For God has destined us not for wrath, but for obtaining salvation through our Lord Jesus Christ, who died for us, so that whether we are awake or asleep we may live with him. Therefore, encourage one another and build each other up, as indeed you are doing.
>
> (1 Thess 5:1–11 [NRSV])

Later, perhaps through the Johannine tradition, the Christian community came to believe that the final fullness was already present in Jesus and already/but not yet fully present in the world through his indwelling Spirit.

Many of the later letters reflect an acceptance of this already/not yet tension or "realized eschatology" and express a greater concern for the continuity of the church and the apostolic tradition, a concern that arose as the apostles and other early witnesses to the life of Jesus died or were put to death.

These letters also speak out against false teachers and false teachings, and clarify misunderstandings that have arisen. They call upon the communities to settle disputes among the people, and exhort people to live in unity and mutual love. They also give moral instructions and express concern for daily affairs in worship and in community life.

Some of these instructions, particularly those concerning women, seem not only inappropriate for our day but also in tension with the role of women in the gospels. This tension, at least for the reader, arises from the span of time between the life of Jesus and the completion of the letters. While the life of Jesus clearly expresses his total inclusion of women within an existing patriarchal society, by the time the letters were completed at the end of the first century, the initial freedom and inclusion that women experienced in Jesus had eroded, and the traditional patriarchal status of women as inferior beings or nonpersons had overtaken the Christian community.

Origins and Collections

The New Testament letters fall into four basic categories. The earliest of these are the letters of Paul to the Christian communities with whom he had a close connection or an intimate relationship. Beginning about 50 C.E., these letters include First Thessalonians, First and Second Corinthians, Galatians, Romans, Philippians, and the letter to Philemon.

A second group of writings arises among the disciples of Paul in the sixties and seventies and builds on his theologies. This group includes the letters to the Colossians and the Ephesians, and Second Thessalonians. It also includes the pastoral letters of First Timothy, Second Timothy, and Titus, and the elegant Letter to the Hebrews. All of these letters are Pauline in thought but are written by later authors who are steeped in Paul's theology and who deepen it and carry it forward into new times and new situations.

These letters of Paul and his disciples comprise fourteen letters and are arranged in the Scriptures according to their length, moving from the longest to the shortest, rather than according to time, theme, or any other consideration. The exception to this is the Letter to the Hebrews, which is quite long and closes the Pauline writings.

The third group, known as the Letters to the Churches, is a collection of four letters of unknown authorship addressed to all the churches. Following the tradition of assigning authorship to significant figures, these letters are given apostolic authenticity in being named First and Second Peter, James, and Jude. They are thought to have emerged near the end of the first century.

The final group of letters arises in the Johannine community. These too are thought to have come into being near the end of the first century and are given apostolic authenticity in being named First, Second, and Third John.

This entire collection of New Testament letters contains twenty-one writings and spans forty or fifty years of time, yet among all these writers, Paul is the only author we know by name. Because of this, he stands out as a giant within this body of writing and within the New Testament itself.

The Pauline Letters

Paul began his life as Saul, a devout Jew and a Pharisee from the city of Tarsus in Asia Minor. We know from the Acts of the Apostles that he was zealous in persecuting Christians and was present as a young man at the stoning of Stephen. We are told that those who stoned Stephen laid their cloaks at the feet of Saul. We also learn from the book of Acts that while on his way to arrest Christians in Damascus, Saul was confronted by a bright light and a great voice. This overpowering religious experience literally stopped Saul in his tracks and turned his life around. As he had been the most zealous persecutor of the Christians, he became the most zealous proclaimer of the Risen Lord Jesus. Through his encounter with the Risen Lord, Saul became Paul (Acts 6:55–9:30).

After this experience on the road to Damascus, Paul spent three years learning his discipleship in Arabia and Damascus and the rest of his life as an itinerant missionary, proclaiming salvation through Jesus Christ. Throughout this missionary activity, Paul made three major journeys, moving deeper and deeper into the heart of the Roman Empire. When at last he reached Rome, he arrived there in chains, a prisoner of the empire. It is thought that he was eventually set free but died there in 67 C.E., during a second imprisonment.

The letters of Paul find their theological center in salvation through Jesus and in freedom from the law. They find their pastoral center in unity and mutual love within the Christian community.

The theological significance of Jesus lay at the heart of the early church and its struggle with the relationship between Judaism and Christianity. Paul responded to this struggle with his own experiential understanding.

For Paul, Jesus is the fulfillment of the Jewish Scriptures, and the Jewish Scriptures are the preparation for the coming of Jesus. Moving from Abraham to Moses to Jesus, Paul looks to Jesus as the universal savior of all humanity. Through this universal grace, all humanity, whether Jew or Greek, slave or free, male or female, finds oneness and salvation in Jesus.

Paul also believed that faith in Jesus frees one from the burden of the law. He came to understand that the external law had been abrogated by a new law that was written on the human heart, and that while the Torah was no longer binding, it served as a historical background for the present moment and unveiled the divine plan for salvation, which found its perfection in Jesus.

For Paul, the Mosaic Law was given by God and was holy, but it could not provide the power and spiritual depth that was needed to obey it. Because of human weakness, it was necessary that this older law be replaced by a new law, and that a new strength and love and Spirit be given so that humanity might remain faithful. Through his own life in the Spirit, Paul came to understand that the Mosaic Law and the covenant with Moses had been profoundly transcended in the New Law and New Covenant of Jesus Christ.

Throughout his letters, Paul also expresses a deep pastoral concern for the members of the Christian community and the communities themselves. Believing that each Christian is called to incarnate the wisdom and love of God in his or her own life, each of his letters supports this wisdom and love with specific instructions on issues and problems in the unique communities.

In these letters, Paul calls all Christians, in family and in community, to mutual love, forgiveness, and harmony. He addresses issues of liturgical order and divisions within the communities. He offers encouragement in persecution, distress, and hardship. He pleads that love, compassion, and fellowship be extended to a runaway slave. Throughout his letters, Paul teaches, preaches, praises, encourages, exhorts, admonishes, and pleads, breathlessly addressing every aspect of the communities' social, political and religious life:

> Let love be sincere; hate what is evil, hold on to what is good; love one another with mutual affection; anticipate one another in showing honor. Do not grow slack in zeal, be fervent in spirit, serve the Lord. Rejoice in hope, endure in affliction, persevere in prayer. Contribute to the needs of the holy ones, exercise hospitality. Bless those who persecute [you], bless and do not curse them. Rejoice with those who rejoice, weep with those who weep. . . . Do not be haughty but associate with the lowly, do not be wise in your own estimation. Do not repay anyone evil for evil, be concerned for what is noble in the sight of all. If possible, on your part, live at peace with all. Do not be conquered by evil but conquer evil with good. . . . Pay to all their dues, taxes to whom taxes are due, toll to whom toll is due, respect to whom respect is due, honor to whom honor is due. Owe nothing to anyone except to love one another for anyone who loves another has fulfilled the law. (Rom 12:9–13:8)

The Writings of the Pauline Disciples

A second group of writings is Pauline in thought, but differs from Paul in literary style and rhetoric. Though these writings have traditionally been understood as the writing of Paul, there is a growing awareness throughout the academic community that some of these letters were written after Paul's death, and that all of these letters are the work of Pauline disciples. Once again, these letters reflect the issues that were alive in the communities to whom they were addressed.

Second Thessalonians, Colossians, and Ephesians

Written perhaps forty to fifty years after First Thessalonians, Second Thessalonians speaks of the Day of the Lord and its attendant issues, suggesting that it is addressed to a community experiencing persecution.

In this letter, the author builds on the authority and earlier writing of Paul, using his teachings in a new context to address a contemporary crisis around end-time. The letters to the Colossians and the Ephesians, once thought to be among the "Captivity Epistles" of the imprisoned Paul, are now coming to be understood as the work of unknown disciples within the Pauline tradition. Drawing on the theology and style of Paul, the author of Colossians moves beyond Paul with a creative theological vision extolling the preeminence of Christ and the wisdom and power of living a profound spiritual life of freedom and mutual love in Christ:

> As God's chosen ones, holy and beloved, clothe yourselves with compassion, kindness, humility, meekness, and patience. Bear with one another and, if one has a complaint against another, forgive each other; just as the Lord has forgiven you, so you also must forgive. Above all, clothe yourselves with love, which binds everything together in perfect harmony. And let the peace of Christ rule in your hearts, to which indeed you were called in the one body. And be thankful. (Col 3:12–15 [NRSV])

Drawing on both Paul and the Letter to the Colossians, the author of Ephesians proclaims the wisdom, power, and mystery of the Cosmic Christ and develops a theology of the Church as the Body of Christ. For the author of Ephesians, the Church is holy through the redeeming blood of Christ, catholic in its universal inclusion of Gentile and Jew, and apostolic in its beginning among the early community of disciples. For this author, the community of believers is the dwelling place of God, where all are made one in the Peace of Christ:

> I, then, a prisoner for the Lord, urge you to live in a manner worthy of the call you have received, with all humility and gentleness, with patience, bearing with one another through love, striving to preserve the unity of the spirit through the bond of peace: one body and one Spirit, as you were also called to the one hope of your call; one Lord, one faith, one baptism; one God . . . of all, who is over all and through all and in all. (Eph 4:1–6)

The Letter to the Hebrews

The elegant Letter to the Hebrews is thought to be a written homily addressed to a weary and disheartened community of Jewish Christians that had been exiled from Jerusalem. Remembering the displacement and the desert journey of the early Hebrews, the unknown author draws on the wilderness motifs of the Hebrew Scriptures and images the present community as the new Hebrews on a pilgrimage toward a heavenly home and a sabbath rest. Focusing on the person and mission of Jesus, on his priesthood and sacrificial death, the unknown author offers encouragement and calls the community to persevere in faith and hope.

The author of this letter also proclaims the preeminence of Jesus. He perceives Jesus as the great high priest of the order of Melchizedek, as the compassionate one who is greater that the angels, who was beset by weakness and made perfect through suffering, who consecrates the children of God and leads them on to their heavenly home. The author calls the community to grow in mature Christian faith and to inherit the promises made to Abraham. Like Paul, he understands Jesus as the mediator of the new covenant and sees the sacrifice of Jesus replacing the sacrifice of goats and calves. He echoes Paul in proclaiming that the old covenant and the old law have been replaced by the new covenant and the new law, which is written on the heart. He calls the community to be faithful as their ancestors were faithful and to enter into the consuming fire of God's love:

> Faith is the realization of what is hoped for and evidence of things not seen. Because of it the ancients were well attested. . . . Therefore, since we are surrounded by so great a cloud of witnesses, let us rid ourselves of every burden and sin that clings to us and persevere in running the race that lies before us while keeping our eyes fixed on Jesus, the leader and perfecter of faith. . . . So strengthen your drooping hands and your weak knees. Make straight paths for your feet, that what is lame may not be dislocated but healed. Strive for peace with everyone, and for that holiness without which no one will see the Lord. (Heb 11:1–2; 12:1–15)

The Pastoral Epistles

The letters of First and Second Timothy and Titus are pastoral writings that express the self-understanding of a growing and changing church. They seek to bring order into the chaos of diverse or false teachings and teachers and acclaim the present leaders as authentic successors in the apostolic tradition. They address order within the church and continue the advice and moral exhortations of Paul.

The Letters to All Christians

The letters of James, First and Second Peter, and Jude speak traditional exhortations and moral teachings to the broader Christian community. By the time of their writing, in the later years of the first century, the Christian communities were being recognized as one church, and there was a common treasury of wisdom and moral teaching. This teaching addressed perseverance in trials, living in mixed communities, respect for civil authority, conflicting teachings, household affairs, humility, patience, holiness, prayer, faith, good works, and the delay of the Parousia. These letters portray a church struggling with its heritage, its contemporary pluralism, and its attempts to bring the wisdom of the past into a different time and cultural setting.

The Johannine Letters

The Johannine Letters, written by an unknown disciple in the last years of the first century, find their way into the Christian Scriptures through their similarity to the Gospel of John and their association with the community of the Beloved Disciple in Ephesus. The first of these is the longest and most significant, stressing faith and love in Jesus and love for one another within the Christian community. It is thought that it was written to clarify the gospel and to correct errors that were emerging in the teaching of the gospel. Bearing some affinity to the Gnostic tradition and the teachings of the Essenes, the gospel had expressed a very high Christology in which Jesus comes from above and returns to the glory he had from the beginning. The letter balances this imagery. As the gospel opens with a soaring hymn in remembrance of the Incarnate Word, the letter grounds this lofty portrayal of Jesus by opening with a simple hymn affirming his humanity:

> What was from the beginning,
> what we have heard,
> what we have seen with our eyes,
> what we have looked upon
> and touched with our hands
> concerns the Word of life—
> for the life was made visible;
> we have seen it and testify to it
> and proclaim to you the eternal life
> that was made visible to us—

what we have seen and heard
we proclaim now to you,
so that you too may have fellowship with us;
for our fellowship is with God
and with God's Son, Jesus Christ.
We are writing this so that our joy may be complete.

(1 Jn 1:1–4)

ECO-SPIRITUAL REFLECTIONS

The Hymns of the Universe

The New Testament hymns are rooted not only in the musical traditions of Israel, but in the core experience of being human and the existential fabric of the universe. Human existence has always been filled with the sounds of drums and chants, and creation has been resplendent with the song of birds and whales and wind. Within this larger tradition, the tribes of Moses played their instruments in their journey through the wilderness of Sinai, and Joshua is said to have brought down the walls of Jericho with the sounds of trumpets. In the stories of kingship, David is remembered as a skilled musician and composer, who played and danced before the Lord, and David's psalms became the core of the temple psalter and the Book of Psalms.[1] Liturgical scholar Michael Witczak writes of this musical tradition within the Scriptures:

> To play music before the Lord came to be an essential part of the worship of Israel. We learn of the levitical orchestra and choir who played in the temple. We hear also of the music that accompanied the reading of the Law when Nehemiah and Ezra regrouped the people after the Exile.
>
> This tradition of music before the Lord is an essential one in Christianity. Jesus and his disciples sang the Great Hallel as they set out to the Garden of Olives. Paul enjoined the Colossians and Ephesians to engage in "psalms, hymns, and

inspired songs." The letters of the New Testament and the Book of Revelation are full of the hymns to Christ that the early community sang when they gathered.

("Music and Saint Francis Seminary," *Salesianum,* Spring/Summer 1993, p. 4)

These New Testament hymns, which center on the person and work of Jesus, arise amid the many hymns of the universe and initiate a new theology that celebrates the emergence of Jesus as the Cosmic Christ. The earliest expression of this new awareness appears in Saint Paul's letter to the Philippians where we see Jesus as having been enthroned in the heavens through his willingness to lay down his life in humility, fidelity, and love:

Have among yourselves the same attitude that is also yours in Christ Jesus,
Who, though he was in the form of God,
did not deem equality with God something to be grasped.
Rather, he emptied himself,
taking the form of a slave,
coming in human likeness;
and found in human appearance,
he humbled himself,
being obedient to death,
even death on a cross.
Because of this, God greatly exalted him
and bestowed on him the name
that is above every name,
that at the name of Jesus
every knee should bend,
of those in heaven and on earth and under the earth,
and every tongue confess
Jesus Christ is Lord,
to the glory of God the Father.

(Phil 2:5–11)

This new understanding of the sovereignty of Jesus continues in the letters to the Hebrews, the Colossians, and the Ephesians, and in the prologue of the Gospel of John. In these hymns, we experience Jesus as the preexistent Eternal One who shares divinity with the transcendent God, who comes forth from God as Wisdom and Word in the humanity of Jesus, and who lays down his life and is raised up, ascending to the heavens and gathering all things to himself.

Biblical Roots

This celebration of Jesus as the Cosmic Christ came easily to the early Christian community. The ancient Near East had long been immersed in rich cosmologies, and among the Israelites God was always present to watch over the people, angels were always available to relay messages, and creation itself was always alive and full of praise.

In the covenant tradition, God appeared to Abraham and Jacob in angelic form, and to Moses in fire and storm. Among nonhuman creation, mountains and rivers, stones and trees, were perceived as conscious beings and called to bear witness in covenants and contracts. In the prophetic tradition, holy men and women were attentive to the voice of God, and Isaiah and Ezekiel had visions of angels and heavenly thrones. In the Wisdom Literature, the mysterious and elusive Spirit of Wisdom moved freely in the heights and in the depths, and her existence preceded time and light.[2] In the worshiping community, the whole of creation was called to participate in hymns of praise:

> Praise the LORD from the heavens,
> praise him in the heights;
> Praise him, all you his angels,
> praise him, all you his hosts.
> Praise him, sun and moon;
> praise him, all you shining stars.

Praise him, you highest heavens,
and you waters above the heavens. . . .

Praise the LORD from the earth,
you sea monsters and all depths;
Fire and hail, snow and mist,
storm winds that fulfill his word;
You mountains and all you hills,
you fruit trees and all you cedars;
You wild beasts and all tame animals,
you creeping things and you winged fowl.

Let the kings of the earth and all peoples,
the princes and all the judges of the earth,
Young men too, and maidens,
old men and boys,
Praise the name of the LORD.
for his name alone is exalted . . .

(Ps 148:1–13)

Rooted in the Wisdom Literature and the Psalms of the Hebrew Scriptures, which celebrate creation as "a communion of subjects, rather than a collection of objects,"[3] the Christological Hymns of the Christian Scriptures celebrate creation as a single celebratory whole that is created, sustained, and continually moving toward its final fullness in the Cosmic Christ:

He is the image of the invisible God,
the first born of all creation.
For in him were created all things in heaven and on earth,
the visible and the invisible,
whether thrones or dominions or principalities or powers;
all things were created through him and for him.
He is before all things,
and in him all things hold together.
He is the head of the body, the church.

He is the beginning, the first born from the dead,
that in all things he himself might be preeminent.
For in him all the fullness was pleased to dwell,
and through him to reconcile all things for him,
making peace by the blood of his cross,
[through him], whether those on earth or those in heaven.

(Col 1:15–20)

Post-Biblical Understandings

This understanding of Jesus as the Christ of the cosmos, which was rooted in the cosmological theology of the Hebrew Scriptures and came into being in the New Testament Letters, flowered after New Testament times among the early "Fathers" of the Church. These men gathered in two great schools of thought in response to the questions and struggles of the Christian community.

The first of these schools was the Semitic school of Antioch in Syria where Ignatius, Polycarp, Irenaeus, and Athanasius served as teachers. This school followed the biblical traditions of Paul and John and looked to the risen Jesus in a dynamic and existential way that embraced both immanence and transcendence. It also fostered a mystical, Spirit-filled spirituality and remained open to the feminine in God.

The second school was the school of Alexandria in Egypt, led by Clement, Cyril, and Origen. This school was influenced by Neoplatonism and had a more philosophical and systemic perspective. It looked to the human Jesus as the teacher and model for right living and to the risen Jesus as the Logos or the reason and mind of the universe.

In this school, the movement from preexistent Word to the human Jesus to the Cosmic Christ, first noted by Paul in his letter to the Philippians, was carried forward by Origen in the third century. While the Christian Church of that time was struggling with the theology of the incarnation and the relation of the human and the divine in Jesus, Origen was struggling with the cosmic dimension of Jesus and the possibility of a plurality of inhabited worlds in the all-inclusiveness of the Alpha and the Omega of the Book of Revelation:

> God the Logos is the Alpha, the beginning and cause of all things, the one who is first not in time but in honor . . . Let it be said that since he provides an end for the things created from him . . . he is the Omega at the consummation of the ages. He is first and then he is last, not in relation to time, but because he provides a beginning and an end. Here are understood the extremities of the letters, which are the beginning and the end and include the others in between.
>
> (Scholia of Origen on the Apocalypse.)[4]

After the Council of Nicea in 325 C.E., a third school of thought, influenced by Basil, Gregory of Nazianzus and Gregory of Nyssa, gained prominence and brought forth a more monastic, mystical and personalist spirituality.

Though all these theologies looked to and sang praise to Jesus as the Christ of the cosmos, with the passing of time, the cosmos came to function primarily as the background for the drama of human salvation. The natural world, which had been included in the creation theology of the Hebrew Scriptures and the spirituality and teaching of Jesus, lost its recognition as a bearer of the Divine.

While the influence of Platonic philosophy was undoubtedly responsible for some loss of joy and wonder in the natural world, it is also possible that the loss of the natural world itself was responsible for the more anthropocentric theologies. Unlike the Hebrew Scriptures and the spirituality of Jesus, which grew in the earthy communities of Israel, these early post-biblical theologies developed in the city with its concern for the affairs of the city, as well as the state and the empire. As spiritualities arise from their setting in life, it is quite possible that this lack of immersion in the world of nature diminished the awareness of the natural world as the resplendent bearer of Divine Presence.

Within these philosophical theologies, the Cosmic Christ came into our contemporary twentieth-century thought in the theology of paleontologist Teilhard de Chardin. Teilhard understood the Cosmic Christ as the Omega point in an evolving universe that was characterized by an increasing complexification and an increasing rise in consciousness. For Teilhard, Christ, in his third or cosmic aspect, was the organizing principle in this evolving and teleologically oriented universe. Like Origen, Teilhard struggled to reconcile his understanding of the earthly Jesus as the Cosmic Christ with the probability of a plurality of inhabited worlds within the universe. He also perceived, in the Cosmic Christ, the evolving nature of a God in process:

> Since Jesus was born, and grew to his full stature, and died, everything has continued to move forward *because Christ is not yet fully formed:* he has not yet gathered about him the last folds of his robe of flesh and of love which is made up of his faithful followers. The mystical Christ has not yet attained to his full growth; and therefore the same is true of the cosmic Christ. Both of these are simultaneously in the state of being and of becoming; and it is from the prolongation of this process of becoming that all created activity ultimately springs, Christ is the end-point of the evolution, even the *natural* evolution, of all beings; and therefore evolution is holy.
>
> ("*Pensées,*" *Hymn of the Universe,* p. 133)

While Origen and Teilhard understand the Cosmic Christ in a similar way, their understanding arises in differing perspectives. As a Platonist in the Greek school of Alexandria, Origen has a high or "from above" Christology that focuses on being, and the gathering, embracing divinity of the Cosmic Christ. As a scientific evolutionist, Teilhard has a low or "from below" Christology that perceives the inwardness of the Cosmic Christ in both the being and becoming of creation and in the ultimate convergence of all things.

Medieval Mysticism and the Songs of Creation

Between the patristic genius of Origen and the contemporary genius of Teilhard lies the rich, earthy mysticism of the medieval mystics. This creation-centered mysticism arose in the twelfth century in the lush green lands of Central Europe, and celebrated the presence of God in the natural world. It emerged through Saint Francis, amid the splendid mountains and wildflowers of Assisi, and through Hildegard of Bingham in the rich green valleys of the medieval Rhineland. It continued on in the creation-centered mysticism of Mechtild of Magdeburg, Meister Eckhart, Dante Alighiere, and Nicholas of Cusa.[5]

As a troubadour of the Lord Jesus, Francis began his Canticle of the Creatures at San Damiano on the poppy-covered hillside of Assisi:

Most high, all powerful, good Lord,
All praise be yours, all glory, all honor and all blessing....

All praise be yours, my Lord, in all your creatures,
especially Sir Brother Sun who brings the day;
and light you give us through him....

All praise be yours, my Lord, for Sister Moon and the Stars;
in the heavens you have made them, bright and precious and fair.

All praise be yours, my Lord, for Brother Wind and the Air,
and fair and stormy and every kind of weather
by which you nourish everything you have made.

All praise be yours, my Lord, for Sister Water;
she is so useful and lowly, so precious and pure.

All praise be yours, my Lord, for Brother Fire
by whom you brighten the night.
How beautiful he is, how gay, robust and strong!

All praise be yours, my Lord, for Sister Earth, our mother
who feeds us, rules us

and produces all sorts of fruit and colored flower and
herbs....

("Canticle of the Creatures,"
St. Francis at Prayer, pp. 42–43)

In her spiritual awakening, Mechtild experienced a childlike innocence and wonderment in discovering the Divine Presence in the heart of all creation:

The day of my spiritual awakening
was the day I saw
and knew I saw
all things in God
and God
in all things.

(*Meditations with Mechtild
of Magdeburg,* p. 42

Meister Eckhart, who spoke from his own experience of God, taught ordinary people to trust in the wisdom of their own hearts and to find God in all things:

Apprehend God in all things,
for God is in all things....
Every single creature is full of God
and is a book of God....
Every creature is a word of God....
If I spent enough time with the tiniest creature—
even a caterpillar—
I would never have to prepare another sermon. So full of
God is every creature.

(*Meditations with Meister Eckhart,* p. 14)

All these writings affirm the Cosmic Christ as the one in whom all things came into being and in whom all things are gathered and brought to a final fullness. They reveal that the songs of Francis, the poems of Mechtild and Dante, the music and painting of Hildegard, the homilies of Eckhart, and the musings and strivings of Origen and Teilhard add the human voice to a universe already alive with the exuberant hymn of creation.

This hymn, which began in the flaring forth of the fireball, goes on in the splendor of the night sky, the bursting open of the stars, and the luminous beauty of the moon. It finds expression in the flight of the bird, the song of the whale, the purr of the cat, and the mating dance of the crane. It continues on in the windsong of the willow, the babble of the stream, the thunder of the ocean, and the silent prayer of the snow and stones and the solitary pine at midnight.

The Communities of Creation

The New Testament letters tell us that all things come forth from and continue to be gathered together and lifted up in the Cosmic Christ. They also tell is that we, in the community of the Cosmic Christ, are called to live together in mutual love and service. This simple and elegant reality, which moves us to reverence and awe, finds a personal and immediate grounding in our fidelity to the people with whom we live and in our fidelity to the places in which we live. It calls us to cultivate and care for our small corner of the universe and the sphere in which we make our presence known.

In the Benedictine tradition, the monastic was called, not only to the traditional vows of poverty, chastity, and obedience, but to the vow of stability. The monastic was asked to remain in one place, to make it his or her own, and to cultivate and care for this one small place on the earth. This call required a fidelity to the land — to the mountains and hills, the rivers and streams, the plants and the animals — to the whole of the community's landscape. This concern for fidelity to a place through a vow of stability finds its contemporary expression in the bioregional movement.

The Bioregional Imperative

Bioregionalism is a movement that is centered on the integrity of natural systems, and the awareness that natural systems self-organize and possess their own innate wisdom. It visions the human community within the ecological context of biotic and geologic communities and perceives the need for the human community to explore its oneness with this larger community. Bioregionalism calls for a constructive re-visioning of the human presence in the natural world and seeks a cultural renewal that fosters a mutually enhancing relationship between the human and the nonhuman communities of the earth.

While no one is certain of the origins of bioregionalism, poet Gary Snyder believes it had its origin in local European cultures that sought to maintain their cultural authenticity in the face of the encroaching power of the city. In looking to its purposefulness, Snyder writes:

> The aim of bioregionalism is to help our human cultural, political and social structures harmonize with natural systems. Human systems should be informed by, be aware of, be corrected by natural systems.
>
> Thus, the political side of bioregionalism, for starters, is recognizing that there are real boundaries in the real world which are far more appropriate than arbitrary political boundaries. And that is just one step in learning where we really are and how a place works. Learning "how it works" is an enormous exercise, because we are not taught to think in terms of systems, of society or of nature. (*Turtle Talk,* p. 13)

By definition, a bioregion is a unique community of life-forms that exists within the boundaries of a unique geological territory. Its contours may be determined by the life-forms, the cultures, or the psycho-spiritual qualities of the terrain. Bioregionalist Jim Dodge suggests four criteria for determining the contours of a bioregion.[6]

The first of these is biotic shift. Dodge suggests that as a bioregion consists of a stable community of plant and animal life within a territory, its edges may be discerned when a significant change is observed in the makeup of the biotic community, as well as in the landform, climate, or soil. This becomes evident in areas where forests open out onto grasslands or plains rise up into mountains.

The second setting for a bioregion is a watershed. Within a watershed, all life-forms live interdependently upon the watercourses and develop a unique biotic community of plants, animals, and humans. Any disruption in the integrity of the watershed threatens the entire life-system of the bioregion. This setting and its fragility are evidenced in the Pacific Northwest where excessive logging has caused erosion, the silting up of rivers, damage to aquatic life, and increasing poverty in Native American communities.

A third setting for bioregional integrity is our awareness of the spirit-of-place. Many people, both primal and contemporary, have been drawn to mountains or other geological formations as special bearers of the sacred. Mount Shasta in northern California and Taos Mountain in New Mexico are among the many mountains revered for their sacred character, and the human and nonhuman beings around these mountains draw upon their sacred spirit-of-place.

A fourth means of identifying a bioregion is cultural distinctiveness or the unique ways in which the people of a particular territory express themselves in art, literature, music, custom, costume, and religious ritual. In our contemporary times, this cultural distinctiveness may even be evidenced by cultural diversity. In large cities, such as New York or San Francisco, a multitude of racial, cultural, and ethnic communities live and work together in ways that would be unlikely in other areas.

Listening to the Voices

In authentic bioregionalism, every being needs to be loved, respected, and listened to with care. Decisions concerning the well-being of a community need to be based on the concerns of all beings and life-systems in the community, and arrived at through consensus. Drawing on the deep ecology perspective, bioregionalism invites us to sensitive listening and affective bonding with the community of creation. It invites us to attain at-one-ment with it, to enter so deeply into the interiority and subjectivity of landforms and life-forms as to be at one with them. It calls us to participate in their suffering and joy, and to know their needs and their longings.

From this deep place of compassion and love, we may truly think like a mountain and know the patience of a stone. From this deep place of compassion and love, we become capable and worthy of speaking for these landforms and life-forms in the community of creation and discover a wisdom and a way to participate in the reinhabitation of the land.

Reinhabiting the Land

While many places on the earth are being assaulted and cry out for preservation, other places have been worn out through overuse, abuse, or exploitation and cry out for rehabilitation or restoration. These issues of preservation, rehabilitation, and restoration are not a simple matter of environmental concern. They involve a whole new way of perceiving the world, and a radical altering of our social, political, economic, and religious mentalities. As inheritors of a Western consciousness, we have been encultured to perceive nature as "other," as cruel and violent and soulless, as red in tooth and claw. Murray Bookchin looks to the need for a new ecological sensibility that permeates every aspect of our cultural consciousness:

> There is a need for a new sensibility, a new feeling of care and of love for all forms of life, a feeling or responsibility, a feeling of attunement with the natural world that we are destroying today. It's terribly important that every environmental issue be examined in the light of its social causes. But

I think, too, that this involves a spiritual revolution in our outlook toward each other and toward the natural world. We need a sense of our place in the natural world such that we, as products of nature, act in the service of natural evolution as well as social evolution....

We have developed a culture in which we've said that, to move ahead in history, to advance, the whole goal is above all to grow, grow, grow, grow, grow. If you don't grow, you die, because your rival will devour you. So that, now, the grow-or-die mentality is having disastrous ecological effects. It means cutting down the forests, turning them into pulp to preach a gospel of more consumption; it's making tawdry, cheap, rotten goods; it's being indifferent to the soil, trying to get as much out of it as possible, turning it into a chemical sponge rather that real earth; it's polluting air and water, in the name of industrial growth, with enormous effects as far as the climate and the ozone layer are concerned.

(*Turtle Talk,* pp. 126–27)

This new sensibility is already alive and well in many people and many places in our world, and addresses the ecological tasks of preservation, rehabilitation, and restoration. Preservation is directed toward authentic wilderness areas and areas that are becoming worn through abuse or overuse. The cry to preserve Yosemite National Park from further degradation and the political struggle to preserve open lands in suburban areas are among these works of preservation.

Rehabilitation seeks to recover lands that have experienced extreme degradation. It seeks to rebuild them in ways that are sustainable and support life, even though they will never again experience their original integrity. The efforts of San Francisco's Planet Drum Society to remove sidewalks and to plant vegetation in median strips is a work of rehabilitation. The efforts to reduce erosion at clear-cutting sites by building terraces and planting fruit trees and fast-growing vegetation is another work of rehabilitation.

Restoration is the task of returning a region to its original integrity. Left to themselves, many lands will recover over hundreds or thousands of years. If the entire ecosystem or the biosphere is not damaged, a tropical forest that has been slashed and burned will recover in about a hundred years. In the same situation, a tropical forest that has been cleared by a bulldozer will require a thousand years. Humans can support nature in this recovery, and we can seek its restoration, though this can never be fully accomplished. Restoration requires painstaking research, intense work, and a great deal of time. The Curtis Prairie Project at the University of Wisconsin-Madison Arboretum is one such venture. Conceived by Aldo Leopold and begun in the mid 1930s, this restored prairie is now very near to native prairie land. Other restoration projects focus on wetlands, forests, and small areas within our larger cities. Cities, themselves, cannot be restored.

As we come to the end of this millennium, all our great cities are unholy because they have crowded out the holy of the natural world. The concrete has crowded out the green plants. The lighting has crowded out the night sky. The poor and the young gather on dreary streets. The wild animals lie splattered on the freeways. Industrialization, overpopulation, congestion, excessive consumption, crime, toxic spills, and the exhaust of cars and diesel-powered trucks and buses have made the city uninhabitable. Cities also deplete eco-systems. Unable to support themselves, they drain vast resources from surrounding areas and return their wastes and toxins to these areas. Cities turn wilderness into wasteland, and destroy species, habitat, and the human as an eco-spiritual being and a child of the universe.

On Coming Up from Eden

We have all come up from wild places. They have refreshed us, sustained us, and given us life. Wild places hold us in the embrace of creation, and give us a sense of our personal and cosmic authenticity. They ground us in time and space, in the here and now, and they ground us in timelessness and Eternal Presence. As bearers of the Divine, and as that which we can see with our eyes, touch with our hands, hear and taste and smell, wild places participate in and reveal the sacred. They immerse us in holiness made manifest.

In coming into the wilderness of Walden Pond, Henry David Thoreau wrote:

> What I have been preparing to say is this, in wilderness is the preservation of the world.... Life consists of wildness. The most alive is the wildest. Not yet subdued to man, its presence refreshes him.... When I would re-create myself, I seek the darkest wood, the thickest and most interminable and to the citizen, most dismal, swamp. I enter as a sacred place, a *Sanctum Sanctorum.* There is the strength, the marrow, of Nature. In short, all good things are wild and free.
>
> ("Walking," from *Excursions,* 1863.)

At the turn of the last century, naturalist John Muir struggled to save wild places in his efforts to preserve Yosemite and prevent the damming of the Hetch-Hetchy Valley:

> ...Everybody needs beauty as well as bread, places to play in and pray in, where Nature may heal and cheer and give strength to body and soul.... Nevertheless, like everything else worthwhile, however sacred and precious and well-guarded, they have always been subject to attack, mostly by despoiling gains-seekers... supervisors, lumbermen, cattlemen, farmers, eagerly trying to make everything dollarable, often thinly disguised in smiling philanthropy, calling pocket-filling plunder "Utilization of beneficent natural resources, that man and beast may be fed and the dear Nation grow great." Thus long ago a lot of enterprising merchants made part of the Jerusalem temple into a place of business instead of a place of prayer, changing money, buying and selling cattle and sheep and doves.
>
> ("The Hetch-Hetchy Valley," *Sierra Club Bulletin,* January 1908, pp. 256–57)

Thirty years later, conservationist Aldo Leopold wrote his lament over the loss of wilderness, of things wild and free, from his Sand County farm in South Central Wisconsin:

> WILDERNESS is the raw material out of which man has hammered the artifact called civilization.
>
> Wilderness was never a homogeneous raw material. It was very diverse, and the resulting artifacts are very diverse. These differences in the end-product are known as cultures. The rich diversity of the world's cultures reflects a corresponding diversity in the wilds that gave them birth....
>
> Many of the diverse wildernesses out of which we have hammered America are already gone....
>
> No living man will see again the long-grass prairie where a sea of prairie flowers lapped at the stirrups of the pioneer. We shall do well to find a forty here and there on which the prairie plants can be kept alive as species. There were a hundred such plants, many of exceptional beauty. Most of them are quite unknown to those who have inherited their domain....
>
> No living man will see again the virgin pineries of the Lake States, or the flatwoods of the coastal plain or the giant hardwoods...
>
> Even if wild spots do survive, what of their fauna? The woodland caribou, the several races of mountain sheep, the pure form of woods buffalo, the barren ground grizzly... Of what use are wild areas destitute of their distinctive faunas?
>
> ("Wilderness," *A Sand County Almanac,* pp. 264–69)

We have all cherished our wild places, our forests and wetlands and fields of flowers. Yet in the brief space of a century, we have moved from Thoreau's deep reverence for wilderness, to Muir's work to preserve it, to Leopold's lament for its destruction. In the years since Leopold's death, we have seen an even greater and more rapid destruction of not only wilderness, but the biosphere itself. The water, the soil, and the atmosphere, which we share with every other human and nonhuman being on the earth, are gravely endangered and losing their ability to support life. Dwindling forests and global warming are changing our weather patterns and intensifying floods. Incessant wars are destroying the land and the people of the land.

This violence to the earth and to the communities of the earth is the greatest challenge of our day. Our present institutions of government, whether they are our nations or the United Nations, are inadequate in this situation. There is, therefore, a great need for each of us and all of us to take it into our own hands, to begin again in our own place, to name our own place as a sacred place, and to cultivate and care for it as if our lives depended on it, because they do. Here, on the threshold of a new millennium, we are called to awaken to the urgency of our present situation, to learn, once again, from the communities of creation, and to begin again the great and holy task of reinhabiting the earth.

Chapter Seven

THE BOOK OF REVELATION

If the prehistory of Genesis invites us to reflect on the beginning of creation, the Revelation of John invites us to reflect on the completion of creation. It looks to a time when all things will be brought into their final fullness by the Alpha and the Omega, the Ancient One who was, who is, and who is to come.[1]

This final book of the Judeo-Christian Scriptures is, perhaps, the most difficult book in the Scriptures for contemporary Christians. Filled with unfamiliar and sometimes bizarre imagery, it is difficult to understand in its symbolism or its structure of meaning, even though they are relevant for every era of history.

The Book of Revelation is apocalyptic literature. It is a crisis or resistance literature that arises in times of persecution and oppression and offers encouragement, hope, and consolation when it is hard to go on, when people are wavering and ready to give in. Speaking to the present social and political time of difficulty, it calls the community to fidelity and endurance and points to an eschatological future when God's justice will fully encompass creation and God will be all in all.

Apocalyptic literature is also mythic and poetic literature. It speaks to the intuitive and perceptive realms of the heart and reveals multiple levels of meaning. It does not express a literal reality, but leads us to understand the meaning of things. Arising through the symbolic images of visions and dreams, it draws upon the symbolic imagery embedded in the writer's tradition and is readily understood by its intended audience. Therefore, while the meaning and symbolism of this writing is difficult for us to understand in the present time, it was quite intelligible to the early community of Christians.

Apocalyptic literature is characterized by a dualistic perspective that sets good against evil and "them" against "us." It struggles with the cosmic shadow and the collective human shadow and makes a tremendous cry for justice and the saving intervention of God. Its anguish covers over, at least for a time, any call to love the enemy or overcome evil with good. It is perhaps for this reason that while apocalyptic writings flourished in the first century C.E., only one of these writings found a place in the Christian Scriptures.

The Origins of Apocalyptic Writing

The apocalyptic writings emerge during the Babylonian exile in the visions of Ezekiel, and are carried forward in the writings of the post-exilic prophets, particularly Zechariah. They reach a fully developed genre in the second century B.C.E. in the Book of Daniel. While the apocalyptic expression arises within the prophetic tradition, it seems to supplant it at times. It differs from the prophetic form in that the prophetic form is associated with words and visions and is usually communicated through speech. The apocalyptic form comes or is expressed through vision and is usually communicated in writing. In addition, the prophetic form usually looks to "our" infidelity and calls "us" to justice, while the apocalyptic looks to the wickedness of "others" and calls for justice to be meted out against "them." While classical prophecy began to decline in the post-exilic Persian period, apocalyptic writing rose in this time and flowered between 200 B.C.E. and 100 C.E. In the time of the New Testament writings, both prophetic and apocalyptic forms were prevalent, and prophetic-apocalyptic circles existed.

The Author and the Community

The author of the Book of Revelation identifies himself as John and tells us that he has been exiled to the island of Patmos because of his faith. Though the church of the second century C.E. understood this John to be the evangelist of the fourth gospel, later theologians have dismissed this possibility through the internal evidence of the writing. The writing has some linguistic and theological affinities for the gospel, yet its language, style, and other theological issues indicate a different source.

The author seems, however, to be a prominent and well-respected leader in the early church. His exile to the penal colony of Patmos and the authority of his letters to the churches establish him as a very great authority in the Christian community and, most likely, a member of a prophetic-apocalyptic circle familiar with both the Johannine and Pauline traditions.

The book itself is thought to be the reworking of several earlier apocalyptic visions or writings by the author. Like other apocalyptic writings, it borrows from and builds on the prophets and earlier apocalyptic works, drawing imagery from Isaiah, Jeremiah, Ezekiel, Daniel, and Zechariah, as well as the historical experiences of the Jewish people. It addresses the disparity between the expectations of Christians and the reality of the Roman world and looks to the time of judgment and the fullness of the reign of God as experiences of exodus and liberation.

Because of its profound concern with issues of persecution, this book was most likely written around 95 C.E. during the brutal persecutions of Domitian in Rome. The communities of John in Asia Minor were not experiencing the direct persecution and violent martyrdom of the Christian communities in Rome, but they were experiencing imprisonment, exile, and harassment, as well as the social and political coercion of the Roman culture. It was an era of emperor worship, when even the insane Caligula, the mindless Nero, and the brutal Domitian were honored as gods, and to refuse to cultivate their favor, even by token worship, placed them outside the sphere of social, political, and economic inclusion, and inside the sphere of poverty, oppression, and harassment.

While some Christian communities in the empire believed it was acceptable to participate in these cultic activities because they were civil, political, and even recreational, John perceived a darker and more insidious power behind the imperial power. He passionately believed that the Ancient One and the Lamb were the ones to be worshiped and called the people out beyond the oppression of exclusion, exhorting them to stand against it, to endure, and to remain faithful.

Elisabeth Schüssler Fiorenza understands this writing as a countercultural exhortation, directed against the injustice and dehumanization of the sociopolitical situation:

> It [The Book of Revelation] seeks to alienate the audience from the symbolic persuasion of the imperial cult, to help them overcome their fear so that they not only can decide for the worship and power of God and against that of the emperor but also to stake their lives on this decision. The Book of Rev. is written for those "who hunger and thirst for justice" in a socio-political situation that is characterized by injustice, suffering, and dehumanizing power. In ever new contrasting images the rhetoric of Rev. elaborates the opposition between the life-giving power of God and the death-dealing power of Rome without falling prey to a total metaphysical or ethical dualism.
>
> (*The Book of Revelation: Justice and Judgement,* p. 6)

The Meaning and the Message

The Book of Revelation is not simply an apocalyptic writing. It is also an eschatological writing. The author looks to the present situation in the Christian community and draws out a prophetic understanding that reinterprets the Second Coming of Jesus and forms a bridge between the present and the future, between the "already" and the "not yet" of realized eschatology.

The book begins with a greeting, which identifies the author as a servant and witness of Jesus Christ, and an address, which identifies it as a circular letter to be shared among the Christian communities. This prologue introduces the three basic areas of the book: the Letters to the Churches; the coming time of judgment; and the final fullness of the reign of God:

> The revelation of Jesus Christ, which God gave to him, to show his servants what must happen soon. He made it known by sending his angel to his servant John, who gives witness to the word of God and to the testimony of Jesus Christ by reporting what he saw. Blessed is the one who reads aloud and blessed are those who listen to this prophetic message and heed what is written in it, for the appointed time is near. (Rev 1:1–3)

The first part of the book, the Letters to the Churches, is addressed to seven churches, and as seven is the number of completion, the writing is understood as being addressed to the whole church. Each of the letters follows a basic form. It is written to the "angel" or guardian of the church and begins with appellative praise identifying Jesus as the one who speaks to the community. In his address, the Risen Jesus expresses his knowledge of the goodness and fidelity of the community and his knowledge of their weaknesses and failures. This is followed by a call to repentance and the announcement that he is coming soon. The exceptions to this form are the letters to the churches of Smyrna and Philadelphia, in which there is no mention of failure and no call to repentance. Each letter closes with a promise of salvation and blessing to those who have remained faithful:

> To the angel of the church in Philadelphia, write this:
>
> " 'The holy one, the true,
> who holds the key of David,

> who opens and no one shall close,
> who closes and no one shall open,'

says this:

> "'"I know your works (behold, I have left an open door before you, which no one can close). You have limited strength, and yet you have kept my word and have not denied my name. . . . Because you have kept my message of endurance, I will keep you safe in the time of trial that is going to come to the whole world to test the inhabitants of the earth. I am coming quickly. Hold fast to what you have, so that no one may take your crown. The victor I will make into a pillar in the temple of my God . . . On him I will inscribe the name of my God. . . ."'"
>
> (Rev 3:7–12)

The central section of the book contains a skillful interweaving of visions in which cosmic rule is given to Jesus as the Lamb who was slain (4,5), and a cosmic judgment is given against the unjust nations (6–20). This section is filled with thunder and lightning, angels and trumpets, elders and martyrs, destruction and plagues. It portrays a cosmic liturgy of enthronement and a cosmic courtroom drama where the cries of the martyrs are heard and justice and judgments are given against both the empire and the cosmic shadow that stands behind the empire and the present historical struggle:

> When he [the Lamb] broke open the fifth seal, I saw underneath the altar the souls of those who had been slaughtered because of the witness they bore to the word of God. They cried out in a loud voice, "How long will it be, holy and true master, before you sit in judgement and avenge our blood on the inhabitants of the earth?" Each of them was given a white robe, and they were told to be patient a little while longer . . .
>
> (Rev 6:9–11)

The final section of the book envisions the new heaven and new earth of the eschatological future where justice will reign and God will be all in all (21:1–22:5). The beginning and ending of the book, representing the present and the future, enclose the central apocalyptic section of the book and express the encompassing and protecting presence of Jesus across time.

The book closes with an epilogue affirming these revelations and promising blessing to those who hear and attend its message. Finally, Jesus announces, once again, that he is coming soon, and John responds with the cry of the early church, "Come, Lord Jesus" (22:20).

The Vengeance Problem

The Book of Revelation did not come into the Christian canon with ease. Though early theologians, such as Justin, Irenaeus, and Tertullian, accepted this writing as the work of the apostle John, the churches of Syria, Cappadocia, and Palestine did not accept it into their canons until the fifth century. While early scholars struggled with its authenticity as a Johannine writing, contemporary scholars struggle with the violence of its apocalyptic images and the dualistic perspective that looks to the destruction of the enemy. These scholars experience the book as antithetical to the gospel of Jesus and offensive to both moral and religious sensitivities.

William Barclay struggles with this issue in the call to rejoice over the judgment against Babylon: "Rejoice over her . . . For God has judged your case against her" (8:20). While a simple reading would reveal this as a call to rejoice over the victory of justice and the winning of a lawsuit, Barclay perceives it as a tension between the call of the gospel and the human experience of suffering and rage:

> It is not the more excellent way that Jesus taught. . . . It may be that we are far from the Christian doctrine of forgiveness, but we are very close to the beating of the human heart.
>
> (*The Revelation of John,* Vol. 2, pp. 213, 195)

Adela Yarbro Collins understands these images as a perennial and somewhat effective method of releasing tension around a perceived crisis. Looking to Rollo May's understanding that anger and violence may be life-giving for individuals who live at a subhuman level and lack self-consciousness, assertiveness, and dignity, Collins sees its validity as a partial and imperfect means of raising consciousness and engendering hope:

> These considerations suggest the political stance and conflictual tone of Revelation served the valid purpose of raising the consciousness of certain marginal and frustrated early Christians. Their commitment to a hope for the future that involved a transformation of the political and social order was a protest against the injustice of their current situation. The strength of the Apocalypse is the pointed and universal way in which it raises the questions of justice, wealth and power. Revelation serves the value of humanization insofar as it insists that the marginal, the relatively poor and powerless, must assert themselves to achieve their full humanity and dignity.
>
> (*Crisis and Catharsis: The Power of the Apocalypse,* p. 171)

Collins also believes that the concept of justice is culturally shaped by politics, religion, and a multitude of other factors, and that what is cathartic for one person may be inflammatory for another. She believes that, ultimately, the effect of this writing will be determined by the identity of the reader and the reader's situation. She also believes that the dualistic division of humanity in the Apocalypse is a failure in love and that its darkest elements call us to explore our own feelings of envy, resentment, and vengefulness:

> It is a book that expresses anger and resentment and that may elicit violence. Its achievement is ambiguous insofar as aggressive feeling and violence can be destructive as well as constructive. Revelation works against the values of hu-

> manization and love insofar as the achievement of personal dignity involves the degradation of others.
>
> (*Crisis and Catharsis: The Power of the Apocalypse,* pp. 171–72)

Elisabeth Schüssler Fiorenza defends the Apocalyptic cry for vengeance, understanding it as arising from a situation of powerlessness and providing an alternative vision for those who hunger and thirst for justice. For Fiorenza, the apocalyptic vision provides

> the vision of an "alternate world" in order to encourage Christians and to enhance their staying power in the face of persecution and possible execution.
>
> (*The Book of Revelation: Justice and Judgement,* pp. 187–88)

Fiorenza believes, however, that there are no "timeless" readers and that theologians need to assess the impact of the Apocalypse on contemporary believers and engage in an extensive deconstruction and reconstruction of the text.

Paul Minear understands the Book of Revelation as a mythopoeic writing that creates a symbolic universe and evokes multiple levels of meaning. For Minear, "Babylon" represents not only the present historical situation with Rome, but serves as a transhistorical symbol for all the cities and institutions of human power where oppression and suffering abound.[2]

Daniel Berrigan echoes this perception, saying that there is a power behind all the lesser powers, and ultimately, it doesn't matter who is in power. Looking to the patience and impatience of the martyrs beneath the altar (5:9–11), he speaks his own mythopoeia:

> We draw our meaning from the martyrs. We need patience because there will continue to be more of the same before things change. The work of the martyrs is unfinished. The outcry must continue. Time is not finished until the work

of the martyrs is finished. They are the ones who mark time with their blood.

("Following the Peacemaking Jesus,"
Pax Christi Retreat)

ECO-SPIRITUAL REFLECTIONS

Images of End-time

Throughout time, the human community has visioned an end to time. We have visioned an era when oppression and war will cease and justice and peace will fill the earth. In this era, our suffering and tears will be wiped away, and all creation will abide forever as a community of love.

We find these visions and dreams seven to eight hundred years before the coming of Jesus in the messianic prophecies of Isaiah of Jerusalem. Isaiah looked to a time when nations would "beat their swords into plowshares and their spears into pruning hooks" (2:4), and a time of ultimate reconciliation when all beings would live together in peace:

Then the wolf shall be a guest of the lamb,
and the leopard shall lie down with the kid;
The calf and the young lion shall browse together,
with a little child to guide them.
The cow and the bear shall be neighbors,
together their young shall rest . . .
The baby shall play by the cobra's den,
and the child lay his hand on the adder's lair.
There shall be no harm or hurt on all my holy mountain;
for the earth shall be filled with knowledge of the Lord,
as water covers the sea.

(Is 11:6–9)

We find a second perception in which re-creation follows disaster in the lived experience and visions of Ezekiel. For Ezekiel, time came to an end during the fifth century B.C.E., in the destruction of Jerusalem, its temple, and the Davidic kingship. Taken into exile in the year he would have begun his service as a temple priest, Ezekiel contracted a debilitating illness in Babylonia and experienced the sudden death of his wife, who was "the delight of his eyes."[3] For Ezekiel, who carried the anguish of the people and the anguish of God, this experience of destruction was so complete that it required the re-creation of his world and the world of the Israelites. His visions were cosmic, involving the heavens and the earth, storm winds and fire, cherubim and the living God, dry bones, and a new temple[4]:

> The hand of the LORD came upon me, and he led me out in the spirit of the LORD and set me in the center of the plain, which was now filled with bones. . . . Then he said to me: Son of man, these bones are the whole house of Israel. They have been saying, "Our bones are dried up, our hope is lost, and we are cut off." Therefore, prophesy and say to them: Thus says the LORD God: O my people, I will open your graves and have you rise from them, and bring you back to the land of Israel. Then you shall know that I am the LORD, when I open your graves and have you rise from them, O my people! I will put my spirit in you that you may live, and I will settle you upon your land; thus you shall know that I am the LORD. I have promised, and I will do it, says the LORD.
>
> (Ez 37:1–14)

Ezekiel's greatness emerged out of his anguish, and he is remembered not only as the father of apocalyptic literature, but also as the father of Judaism. Through the work of Ezekiel, the people who went into exile as Israelites returned to their homeland as a Jewish community.

Ezekiel's apocalyptic imagery was carried forward by the later prophets Zechariah and Daniel and found its way into the gospels, the New Testament letters, and the Book of Revelation. The gospels of Matthew, Mark, and Luke share the imagery of the great tribulation and the cosmic coming of the Son of Man on the clouds. They tell us there will be signs in the heavens, that the sun will be darkened, the moon will not give its light, and the stars will fall (Matthew 24:15–31; Mark 13:14–27; Luke 17:20–37). Matthew's Gospel also includes the end-time parable of the weeds and the wheat (13:24–30) and the majestic story of the Last Judgment when all people will be judged by their concern for those who are the least in their midst (25:31–46).

While the synoptic gospels carry the disaster tradition of Ezekiel, Isaiah's imagery of ultimate reconciliation continued on in the pneumatic or Spirit tradition of John, a tradition that portrays end-time as beginning with the coming of Jesus and the sending of the Spirit. It is understood as the graced recreation and renewal of all that is. Post-biblical images follow both of these traditions.

Origen, who is considered to be the greatest theologian of the third century, follows the Isaian tradition. Believing in an original integrity and a tragic fall from grace, Origen envisions an ultimate reconciliation in which even the powers of darkness and evil will find salvation. Drawing on the imagery of Paul and finding an echo in C. G. Jung's reflection on the shadow, he believes the destruction of the last enemy is to be understood in terms of reconciliation:

> Stronger than all evil is the Word and the healing power that dwells in the Word.... We think indeed, that the goodness of God through Christ may recall all creatures to one end, even God's enemies being conquered and subdued... Things will not cease to be, but cease to be enemies.... So then, when the end has been restored to the beginning, God will be all in all. (*Contra Celsus; De Principal* 3,6,1–6.)

A second image is that of Teilhard de Chardin. Teilhard believed that matter metamorphosed into psyche, and that a conscious spirit existed throughout the universe. He believed this final fullness would come, not through some great cataclysm outside the ordinary events of life, but through an implosion of freedom and consciousness into the realm of the Spirit, into the Transcendent Other, and that this shift of human consciousness would alter the universe:

> The Parousia will undoubtedly take place when creation has reached the paroxysm of its capacity for union. The unique action of assimilation and synthesis which has been going on since the beginning of time will be revealed at last, and the universal Christ will appear like a flash of lightning amid the clouds of a world which has gradually become sanctified.... At that moment... when Christ shall have emptied of themselves all the powers of creation, he will bring to completion the unification of the universe.... Like an immense wave, Being shall have dominated the agitation of beings. In the midst of a becalmed Ocean whose every drop of water shall be conscious of remaining itself, the extraordinary adventure of the world shall reach its term. The dream of every mystic shall have found its full and legitimate satisfaction. *Erit in omnibus omnia Deus.* (God will be all in all.)
>
> (*The Future of Man,* 1964, 307–8)

Theologians are not the only ones who have reflected on endtime. Artists have painted their pictures, and writers have portrayed their perceptions.

When C. S. Lewis wrote his *Chronicles of Narnia,* he envisioned the powerful and compassionate lion, Aslan, as the image of the Word of God, as the way God would become incarnate in a world inhabited by talking beasts and mythological figures. Raised on the wonderfully imaginative fables of Aesop, and converted to Christianity through the loving, ecstatic theology of Augustine, Lewis created Narnia out of his own joyous contemplation of God and God's creation.

Like the Word of God, who was present to God in the beginning, Aslan, the only child of the Emperor-Beyond-the-Sea, sings

Narnia into being—stars, sun, earth, and living creatures—with the most beautiful song ever heard. Though his home is in his own High Country, Aslan also dwells simply and majestically in the land of Narnia. He delights in being at the center of the dance and plays joyfully among the talking beasts and the mythological creatures. There are enemies in Narnia, however, and there is an end to Narnia. And, like the Gospel of Matthew, in the end, it is all a matter of love.

After many great struggles between the forces of good and the forces of evil in Narnia, there is an apocalyptic battle and a judgment scene that is one of the finest in Christian fantasy. Aslan, standing with his shadow falling on the devastated land, looks into the eyes of every creature. The ones who look upon him with hate or fear pass into his shadow and oblivion. The ones who love him, even though they are awed by him, pass through the great door and into Aslan's own High Country. (*The Chronicles of Narnia: The Last Battle*)

All these images portray our struggle with the concept and experience of evil and with the dark and shadowy aspects of our personal, collective, and cosmic reality. They express our hopes and perceptions that suffering and evil will be overcome, that they will be destroyed or healed, annihilated or integrated.

C. G. Jung has been a pioneer in exploring the shadowy aspects of reality in its many levels of being. Jung's self-experience and work led him to the awareness that beneath our waking consciousness lies the realm of the personal unconsciousness, and beyond this, the realm of the collective unconscious. He also discovered that as we make our personal journey inward toward our deepest and most authentic self, one of the first inner aspects we encounter is that of the shadow.

For Jung, the shadow consists of inner dispositions and potentials that are unrecognized or unaccepted or unlived. They are rejected and repressed for social, cultural, familial, ethnical, or aesthetic reasons, and are usually projected onto "others" who differ in ethnicity, gender, or age. Though these dispositions or potentials of the shadow are often unrecognized, they contain enormous energy and a great capacity for both creativity and destruction. Their creative aspects are usually considered to be "good." Their destructive aspects are usually considered to be "evil."

Jung came to understand that many of these dark aspects that were thought to be "evil" were actually "redeemable." When brought into the light of consciousness and integrated into one's life, these shadowy aspects usually enriched one's life and led to greater energy, creativity, self-realization, and wholeness.

Jung also came to understand that just as an individual possesses, or is possessed by, his or her own shadow, individual cultures contain their unique shadows, and the human collective itself possesses a shadow self. These many shadows are lived and acted out in interacting patterns that influence not only the human but the entire earth community. We see these shadows erupting all around us today in random and domestic violence, in military and corporate wars, in poverty and environmental degradation, in homelessness and massive migration, and in the innumerable situations of human and planetary diminishment.

Like the shadows of the personal realm, these cultural and collective human shadows need to be recognized, acknowledged, and integrated in order to be healed and to come into a new wholeness and transformation. The unrecognized, unloved, and unintegrated shadow can be exceedingly destructive.

The theology of the Christian Scriptures looks to Jesus as the one who has recognized, acknowledged, and integrated the shadowy aspects of the human and cosmic collectives, and brought healing, wholeness, and transformation to the whole of the cosmos. It understands Jesus as having struggled with his own shadow aspects in the desert and in the garden, and as having taken upon himself the collective and cosmic shadow beneath the darkened sky of Calvary. It proclaims that through his taking on the sin of the world, his dying and descent into the darkest realms of being, and his resurrection, Jesus initiated the healing and regathering of all creation and a new community of integrity.

The Christian Scriptures invite us to participate in this re-creation and re-gathering of the community of creation. They suggest that we begin by attending to our own shadows, by withdrawing the projections we have laid upon others, by removing the beam in our own eye before attending to the mote in the eye of another. Only then will we be able to live in justice, compassion, and community, and share in the visions and dreams of the risen Lord Jesus.

Visions and Dreams

The Revelation of John was written to encourage the men and women of the early Christian communities in a time of darkness, confusion, and chaos. It called them to endure and to remain faithful in the midst of their afflictions and offered the hope of a new heaven and a new earth where justice and peace would abound. In holding fast to this hope, it established a sacred center in the new Jerusalem, in the holy city where every tear would be wiped away and death itself would be no more:

> Then I saw a new heaven and a new earth. The former heaven and the former earth had passed away, and the sea [chaos] was no more. I also saw the holy city, a new Jerusalem, coming down out of heaven from God, prepared as a bride adorned for her husband. I heard a loud voice from the throne saying, "Behold, God's dwelling is with the human race. He will dwell with them and they will be his people and God himself will always be with them [as their God]. He will wipe every tear from their eyes, and there shall be no more death or mourning, wailing or pain, [for] the old order has passed away."
>
> The one who sat on the throne said, "Behold, I make all things new. . . . I [am] the Alpha and the Omega, the beginning and the end." (Rev 21:1–6)

In reflecting on the new heaven and the new earth of the Book of Revelation, we might ask ourselves what visions and dreams encourage, inspire, and support us in our present eco-spiritual crisis and enable us to remain faithful and live toward a transformed community of creation. We might ask what visions and dreams help us to organize our social institutions and our cultural forms in ways that lead the human community to a greater congruence with Ultimate Reality and a new and life-giving presence among the communities of earth.

In reflecting on the meaning and message of the Book of Revelation, we might remember that we consciously participate in our own destiny and that our dominant presence on the earth determines the destiny of all the other life-forms on the planet. We might remember that whatever fosters justice, compassion, and inclusion leads to prosperity and life, and whatever violates justice, compassion, and inclusion leads to disaster and death.

In these reflections, we might also listen to the longings of our hearts as they search for a resting place in God and in God's creation. We might listen to the strivings of our spirits as they seek to abide in the Creative Spirit who fills the world with love. For in these deep recesses of our being, we long for a new heaven and a new earth, where we live in harmony with the Divine ordering in creation. We vision and dream toward a time and a place of radical newness, where we will live and move and have our being in union with the creative freedom of God.

In this new heaven and new earth, we will make a radical reaffirmation of the meaning of God, of God as the inner radiance of all things, as the core of their isness and the circumference of their beyondness. We will live out a radiant global spirituality that flows forth from the radiance of God, renewing our awareness of this sacred presence in all of life and in the whole of creation.

In this new heaven and new earth, we will be attentive to the unfolding of the universe in its present moment, and we will live and create our realities in union with that unfolding. We will come to a deep respect for the differentiation in the universe, for the great multiplicity and diversity of individual beings and realities in the universe. We will grow in our awareness of the subjectivity in the created order, of the withinness and psychic integrity of all beings and the primordial sacredness of all realities. We will participate consciously and gratefully in the communion within the universe, in the interdependence and mutual indwelling of all beings, in the love that holds all realities in mutual attraction.

In our new heaven and new earth, we will perceive the earth as a single living organism and understand ourselves as emerging from the earth through the creative intentionality of God. We will return the earth to a radiant community of living beings, developing a new mode of human presence based on the universal principles of justice, compassion, and inclusion. This new human presence will flow forth in harmony with the self-emergence of the universe and permeate our social and political institutions and the vast multiplicity of our cultural forms.

In our new heaven and new earth, the human community will respond in a unified way to the processes and issues of the earth, and individual human activity will be seen in the larger context of the human community and the earth process. Nations will no longer see themselves as the ultimate frame of reference but move beyond nationalism to a primary allegiance to the earth community as a whole. They will not raise arms against any other nation and they will train for war no more. All peoples will dwell in safety and know the comfort and affection of their brothers and sisters. There shall be food and shelter for all peoples, and no one shall watch another die for lack of bread or love.

In our new heaven and new earth, men and women will live together in equality and peace. We will not disparage one another as male or female, Arab or Jew, Asian or African-American. Rather, we will appreciate each other for our diversity and the inherent value of our lives, and we will understand ourselves as one family within the community of creation. Through this profound experience of community, we will come to know that a deep wisdom abides in all religious traditions, and we will live out a deep ecumenism that draws its strength and meaning from the Ultimate Reality of God.

In our new heaven and new earth, we will cultivate a new political will of justice and compassion that respects the rhythms, tensions, and plurality of perspectives that exist within the human community and within the larger community of creation. In this new political will, we will engage representatives for all the beings of the earth. Women and men will listen to the voices of creation and speak for them with authenticity. They will represent bioregions, landforms, and life-forms, and speak for sea and sky, forests and wetlands, wilderness and wild things.

In our new heaven and new earth, our economy will be built on the well-being and self-realization of the earth and the communities of the earth. It will measure prosperity by the quality of life in human communities as well as in biotic communities and in life-systems such as rivers, mountains, oceans, and skies. In this new era, our work will spring from our hearts and give authentic service to the communities of the earth and richness and meaning to our own lives. We will have a great respect for the carrying capacity of the earth and for the ability of the earth to sustain both human and nonhuman beings in their physical-material needs as well as their psycho-spiritual needs. We will have a care-filled concern for the growth or extinction of human and nonhuman populations and for the continuing integrity and radiance of the earth and the communities of the earth.

In our new heaven and our new earth, we will know that it is good to be born and it is good to die, and we will honor both passages as holy. Through this sacred experience of letting go and letting be, we will learn to let go of old forms and systems when they no longer serve the creative movement of God and embrace new and unknown forms and systems in hope and trust. We will learn to see chaos as a transitional phase within the evolutionary process and move peaceably through phases of order and disorder, regathering always around our Sacred Center and moving out of diseasement and into easement, transformation, and healing through attentive listening to this Sacred Center and the present unfolding of the universe.

In our new heaven and new earth, all our cities will be holy. We will safeguard our rivers, uncover our creeks, and let our springs flow free. We will create green spaces, parklands, and wetlands for human and nonhuman creation, and cultivate solar power and wind power, roof gardens, mass transit, and areas for community gatherings. We will establish educational centers that develop the whole person and councils on the arts that nourish the human spirit and embrace the communities of creation. Through poetry, drama, storytelling, music, and art, we will be led into meditation, celebration, ritual, and play around the Creator and the sacred community of creation.

In our new heaven and new earth, we will understand and tell our creation story, our sacred history, in its physical and numinous aspects, in images that evoke wonder and awe and provide the courage and hope we need to sustain us in our journey. We will conceive and give birth to new myths and dreams out of the unfathomable wisdom and knowledge of God, and in union with the unbounded energy of God in this Mysterious One's own self-emergence. We will live so close to this Sacred Center that wisdom and compassion are always with us, and keep with closest custody our listening hearts, attuning them always to the movement of God's Spirit in the creativity of the here and now.

In all these ways, we will enter into and participate more fully in our own humanness and in the creativity of God. Through all these ways, we will move into possibilities and transformations beyond our present capacity for vision or imagination. For we believe that we were made by Love and we are called by Love and will be transformed by Love. And we believe that we will be carried by the creativity of the universe and the Mysterious One who brought it into being, into a future not only unknown but unknowable, that eye has not seen and ear has not heard, into a realm where we will come to know as we are known, to love as we are loved, and where God will be all in all.

We await this radical newness held out to us in hope.

Then I heard every creature in heaven and on earth
and under the earth and in the sea,
and all that is in them,
singing,
"To the one seated on the throne
and to the Lamb
be blessing and honor and glory and might
forever and ever!"

And the four living creatures said,
"Amen!"

(Rev 5:13–14 [NRSV])

END-NOTES

Foreword

1. Our earliest known hominid ancestor, Australopithecus Afarensis, also known as "Lucy," was discovered at Hadar, Ethiopia, in 1974. Vegetarian "Lucy" and her kin probably lived on the grasslands and slept in the trees, 3–5 million years ago. Archeological evidence also suggests that 2–3 million years ago Australopithecus Robustus and Australopithecus Boisei evolved from this early ancestral group. While "Lucy" was quite diminutive, Robustus and Boisei were strong and muscular with massive jaws and teeth. Both Robustus and Boisei disappeared without further evolution. At this same time, a less muscular species with a far greater cranial capacity was evolving through the same hominid line, and *Homo habilis,* the "Handy Man" emerged. Modern humans are thought to have evolved from "Lucy" and her kin through *Homo habilis.*

Introduction

1. This subtitle and the historical material in this section are based on the work of Robert B. Coote in *Early Israel: A New Horizon* and the work of Robert B. and Mary P. Coote in *Power, Politics and the Making of the Bible* (Minneapolis: Fortress Press, 1990).

2. The "sea people," who are thought to have their origins in the Mediterranean lands of Europe, have never been clearly identified. The favored hypotheses identify them as a Greek people from the shores of the Aegean Sea or mariners from Crete.

3. Robert B. Coote, *Early Israel: A New Horizon* (Minneapolis: Fortress Press, 1990), pp. 155–63. Robert B. Coote and Mary P. Coote, *Power, Politics, and the Making of the Bible* (Minneapolis: Fortress Press, 1990), pp. 28–31.

4. Robert B. Coote, *Early Israel: A New Horizon* (Minneapolis: Fortress Press, 1990), pp. 156–58.

5. The Levites were tribal priests who had been faithful to Moses in the desert. They held positions of honor at country shrines and in the towns and villages throughout the countryside. They also served as assistants to the temple priests on a rotating basis. The Deuteronomic tradition developed through this Levitical priesthood. The Aaronites were direct descendants of Aaron, the brother of Moses, and held positions in the temple priesthood. The priestly tradition arose among this Aaronite group. After the exile, the Aaronite priests became the ruling class with the high priest assuming the role of king, while the Levites assisted the Aaronites as scribes, musicians, and liturgists.

Chapter One: The Prehistory of Genesis

1. Lloyd H. Steffen, "In Defense of Dominion," *Environmental Ethics,* Spring 1992, Vol. 14, No. 1, pp. 66–67.

2. Michael Wood, Video Series: *Legacy,* Part One: Iraq.

3. Walter Brueggemann, *Interpretation: Genesis* (Louisville: John Knox Press, 1982), pp. 51–52.

4. Ibid., p. 14. Brueggemann contrasts the setting and intent of the earlier Creation story with its concern for prideful self-assertion and the later Creation story with its concern for hopelessness and despair.

5. Bernhard W. Anderson, "Creation and the Noachic Covenant," *Cry of the Environment: Rebuilding the Christian Creation Tradition* (Santa Fe: Bear and Company, 1984), p. 47.

6. Ibid., p. 47.

7. The image of the rainbow as the warrior's bow, hung up in peace, has its origins with George Mendenhall in *The Tenth Generation* (Baltimore: Johns Hopkins University Press, 1973), pp. 38–48. It has been thoroughly integrated into the milieu of biblical scholarship and is used by both Bernhard W. Anderson and Walter Brueggemann. Its understanding as a unilateral act of disarmament is expressed by Anderson's student Hugh J. Matlack, in an essay in the spring of 1982. Anderson quotes Matlack in his article "Creation and the Noachic Covenant," *Cry of the Environment: Rebuilding the Christian Creation Tradition* (Santa Fe: Bear and Company, 1984), p. 55.

8. Thomas Aquinas, *Summa Theologica,* Part One, Question 47, Article One.

9. Sean McDonagh, *To Care for the Earth: A Call to a New Theology* (Santa Fe: Bear and Company, 1986), pp. 169–70.

10. Lloyd H. Steffen, "In Defense of Dominion," *Environmental Ethics,* Spring 1992, Vol. 14, No. 1, pp. 63–80.

11. Ibid., p. 72.

12. Jürgen Moltmann, *God in Creation* (San Francisco: Harper-Collins, 1985), pp. 276–83.
13. Richard J. Sklba, Lecture on Prophetic Theology, 1978.
14. Jürgen Moltmann, *God in Creation* (San Francisco: Harper-Collins, 1985), pp. 292–96.
15. Saint Augustine, *Confessions,* Book One, Chapter One.

Chapter Two: The Law

1. Exodus 20:8–11; Deuteronomy 5:12–15.
2. Sean McDonagh, *The Greening of the Church* (Maryknoll, N.Y.: Orbis Books, 1990), pp. 74–79.
3. Thomas Berry, *The Dream of the Earth* (San Francisco: Sierra Club Books, 1988, pp. 11–12.
4. Saint Francis of Assisi, "Canticle of the Creatures."
5. "Message of Chief Seattle," in John Seed, Joanna Macy, Pat Fleming, and Arne Naess, *Thinking Like a Mountain* (Philadelphia: New Society Publishers, 1988).
6. The imagery of "birthquakes" arises in the poetry of Norman C. Habel in *Birthquakes* (Philadelphia: Fortress Press, 1974).

Chapter Three: The Prophets

1. Walter Brueggemann, *The Prophetic Imagination* (Philadelphia: Fortress Press, 1978), p. 13.
2. Ibid., pp. 20–24.
3. Abraham Heschel, *The Prophets* (New York: Harper & Row, 1962), Vol. I, p. 26.
4. Ibid., pp. 9–10; 16.
5. Gerd Theissen, *Biblical Faith: An Evolutionary Approach* (Philadelphia: Fortress Press, 1985), p. 58.
6. Abraham Heschel, *The Prophets* (New York: Harper & Row, 1962), Vol. II, pp. 1–11.
7. Gerd Theissen, *Biblical Faith: An Evolutionary Approach* (Philadelphia: Fortress Press, 1985), pp. 105–28.
8. Arthur R. Peacocke, *Creation and the World of Science: The Bampton Lectures* (Oxford: Clarendon Press, 1978), p. 200.
9. Jay B. McDaniel, *Of God and Pelicans: A Theology of Reverence for Life* (Louisville, Kentucky: Westminster/John Knox Press, 1989), pp. 21–31.

Chapter Four: The Wisdom Literature

1. Roland E. Murphy, *Wisdom Literature and the Psalms* (Nashville: Abingdon Press, 1983), pp. 21–25.

2. Brian Swimme and Thomas Berry, *The Universe Story* (San Francisco: HarperSanFrancisco, 1992), p. 17. In speaking of the origins of the universe, most scientists refer to the original energy event as the "Big Bang," a terminology that trivializes the magnificence, mystery, and sacred nature of this event. Swimme and Berry speak of this event as the "primal flaring forth" of the fireball. This poetic imagery recognizes and honors its sacred nature.

3. Ernan McMullin, "Introduction: Evolution and Creation," *Evolution and Creation* (Notre Dame: University of Notre Dame Press, 1985), pp. 11–16.

4. Robert John Russell, "Cosmology, Creation, and Contingency," *Cosmos as Creation* (Nashville: Abingdon Press, 1989), pp. 180–81.

5. Jay B. McDaniel, *Of God and Pelicans: A Theology of Reverence for Life* (Louisville: Westminster/John Knox Press, 1989), p. 37.

6. Thomas Berry, *Befriending the Earth: A Theology of Reconciliation between Humans and the Earth* (Mystic, CT: Twenty-Third Publications, 1991), pp. 15–16.

7. Ibid., p. 16.

8. Within the emerging tradition of Christian feminism, "The Feminine Face of God" has been used to speak of Mary, the Holy Spirit, and the woman of the wisdom tradition. Within this feminist tradition, it has been used in the title or as the title of at least two books. These books are *Woman: Time and Eternity: The Eternal Woman and the Feminine Face of God* by Maria Clara Bingemer (translated by F. McDonagh, publisher n.a.), and *The Feminine Face of God: The Unfolding of the Sacred in Women* by Sherry Ruth Anderson and Patricia Hopkins (New York: Bantam Books, 1991).

9. While scholars have usually referred to the elegant woman of the Wisdom tradition in the courtly imagery of "Lady Wisdom," wisdom scholar Kathleen O'Connor has accorded her a profound strength and integrity in calling her "the Wisdom Woman." As I find O'Connor's naming more suited to my own perceptions and to contemporary feminist thought, I have used O'Connor's naming.

Chapter Five: The Gospels

1. Donald Senior, *Jesus: A Gospel Portrait* (Mahwah, NJ: Paulist Press, 1992), p. 27.

2. The understanding of the gospels as "portraits" has become a common understanding among contemporary biblical scholars.

3. The gospels of Matthew, Mark, and Luke suggest that Jesus made one journey to Jerusalem during his time of ministry. The gospel of John suggests three. As the Gospels are portraits, rather than histories in the modern sense, the number of journeys needs to be understood in terms of the tradition and the theological purposes of the writers.

4. The "Samaritans" of Jesus' time were descendants of the Samaritans who had survived the Assyrian invasion of the Northern Kingdom in 721 B.C.E. As the sixth century B.C.E. descendants of these survivors were not affected by the invasion and destruction of Jerusalem and its temple in 587 B.C.E. or the Babylonian Exile that followed, they were not allowed to assist in the rebuilding of the temple or in temple worship. This prohibition and hostility extended into the first century and the time of Jesus.

5. Marcus Borg, *Jesus: A New Vision* (San Francisco: HarperSanFrancisco, 1987), p. 86.

6. Genesis 12:1–9; Genesis 28:10–22; Exodus 2:23–4:16; Isaiah 6:1–3; Ezekiel 1:1–28.

7. Marcus Borg, *Jesus: A New Vision* (San Francisco: HarperSanFrancisco 1987), pp. 86–87, 157–60.

8. Ibid., p. 131.

9. Martin Luther King, Jr., "An Experiment in Love," *A Testament of Hope* (San Francisco: Harper & Row Publishers, 1986), pp. 16–20.

Chapter Six: The New Testament Letters

1. Exodus 15:1–18; Numbers 10:1–10; Deuteronomy 32:1–44. Joshua 6:1–20. 1 Samuel 18:10; 2 Samuel 6:11–15; 2 Samuel 22:1–57. Many Psalms are attributed to David. Some of these were most likely written by temple musicians "in the spirit of David."

2. Genesis 18:1–33; Genesis 32:23–33; Exodus 19:16–19. Isaiah 6:1–7; Ezekiel 1:1–28. Ecclesiasticus 24:1–9; Proverbs 8:22–31; Wisdom 7:29–30.

3. Thomas Berry, Lecture: "Moments of Grace," 1992.

4. Origen, from "The Text of the Newly Discovered Scholia of Origen on the Apocalypse," corrected by C. H. Turner. *The Journal of Theological Studies,* Vol. 13, pp. 386–97. Quoted in *The Cosmic Christ in Origen and Teilhard de Chardin,* pp. 130–31.

5. Matthew Fox, *The Coming of The Cosmic Christ* (San Francisco: Harper & Row, Publishers, 1988), pp. 109–27.

6. Jim Dodge (Coevolutionary Quarterly 32), pp. 6–12. Summarized by Bill Devall in *Simple in Means, Rich in Ends* (Salt Lake City: Gibbs-Smith, 1988), pp. 60–62.

Chapter Seven: The Book of Revelation

1. Revelation 1:8; 21:6; 22:13.
2. Paul Minear, *I Saw a New Heaven and a New Earth: A Complete New Study and Translation of the Book of Revelation* (Washington, D.C.: Corpus Books, 1968), pp. 233–46.
3. Ezekiel 24:15–18.
4. Ezekiel 1:1–28; 10:1–23; 37:1–14; 40:1–48:35.

SELECTED BIBLIOGRAPHY

Anderson, Bernhard W. "Creation in the Bible," *Cry of the Environment: Rebuilding the Christian Creation Tradition.* Santa Fe: Bear and Company, 1984.

———. "Creation and the Noachic Covenant," *Cry of the Environment: Rebuilding the Christian Creation Tradition.* Santa Fe: Bear and Company, 1984.

Bader, W. *St. Francis on Prayer.* New Rochelle, N.Y.: New City Press, 1990.

Baltazar, Eulalio T. *God Within Process.* Paramus, NJ: Newman Press, 1970.

Berry, Thomas. *Befriending The Earth: A Theology of Reconciliation Between Humans and The Earth.* Mystic CT: Twenty-Third Publications, 1991.

———. *The Dream of the Earth.* San Francisco: Sierra Club Books, 1988.

———. *Thomas Berry and The New Cosmology.* Mystic, CT. Twenty-Third Publications, 1987.

Birch, Charles, and John B. Cobb. *The Liberation of Life.* Cambridge: Cambridge University Press, 1981.

Boat, Lawrence E. *Introduction to Wisdom Literature/Proverbs.* Collegeville Bible Commentary. Collegeville: The Liturgical Press, 1986.

Borg, Marcus J. *Jesus: A New Vision.* San Francisco: HarperSanFrancisco, 1987.

Boring, Eugene M. *Interpretation: Revelation.* Louisville: John Knox Press, 1989.

Brueggemann, Walter. *Interpretation: Genesis.* Louisville: John Knox Press, 1982.

———. *The Prophetic Imagination.* Philadelphia: Fortress Press, 1978.

Cady, Susan, Marion Ronan, and Hal Taussig. *Wisdom's Feast: Sophia in Study and Celebration.* San Francisco: Harper & Row, 1989.

Collins, Adela Yarbro. *Crisis and Catharsis: The Power of the Apocalypse.* Philadelphia: The Westminster Press, 1984.

Coote, Robert B. *Early Israel: A New Horizon*. Minneapolis: Fortress Press, 1990.

Coote, Robert B., and Mary B. Coote. *Power, Politics, and the Making of the Bible*. Minneapolis: Fortress Press, 1990.

Cummings, Charles. *Eco-Spirituality: Towards a Reverence for Life*. New York: Paulist Press, 1991.

Diamond, Irene, and Gloria Feman Orenstein. *Reweaving the World: The Emergence of Ecofeminism*. San Francisco: Sierra Club Books, 1990.

Dowd, Michael. *EarthSpirit: A Handbook for Nurturing an Ecological Christianity*. Mystic, CT: Twenty-Third Publications, 1991.

Duvall, Bill. *Simple in Means, Rich in Ends: Practicing Deep Ecology*. Salt Lake City: Peregrine Smith Books, 1988.

Duvall, Bill, and George Sessions. *Deep Ecology: Living As If Nature Mattered*. Salt Lake City: Peregrine Smith Books, 1985.

Eisler, Riane. *The Chalice and the Blade: Our History, Our Future*. San Francisco: HarperSanFrancisco, 1987.

Fiorenza, Elisabeth Schüssler. *The Book of Revelation: Justice and Judgment*. Philadelphia: Fortress Press, 1985.

Fox, Matthew. *The Coming of the Cosmic Christ*. San Francisco: Harper & Row, 1988.

Fox, Warwick. *Towards a Transpersonal Ecology: Developing New Foundations for Environmentalism*. Boston: Shambhala, 1990.

Gore, Albert. *Earth in the Balance: Ecology and the Human Spirit*. Boston: Houghton Mifflin, 1992.

Granberg-Michaelson, Wesley. *Tending the Garden: Essays on the Gospel and the Earth*. Grand Rapids: William B. Eerdmans Publishing Company, 1987.

Heschel, Abraham J. *The Prophets,* Vols. I & II. New York: Harper Torchbooks, 1962.

Jesudasan, Ignatius, S.J. *A Gandhian Theology of Liberation*. Maryknoll, NY: Orbis Books, 1984.

King, Martin Luther, Jr. *A Testament of Hope: The Essential Writings of Martin Luther King, Jr.,* ed. James M. Washington. San Francisco: Harper & Row, 1986.

Kodell, Jerome, O.S.B. *The Gospel According to Luke*. Collegeville: The Liturgical Press, 1989.

Kowalski, Gary A. *The Souls of Animals*. Walpole, NH: Stillpoint, 1991.

Krodel, Gerhard A. *Revelation: Augsburg Commentary on the New Testament*. Minneapolis: Augsburg Publishing House, 1989.

Kurtz, William S., S.J. *The Acts of the Apostles*. Collegeville: The Liturgical Press, 1991.

Leopold, Aldo. *A Sand County Almanac.* New York: Oxford Press, 1949.

Lilburne, Geoffrey R. *A Sense of Place: A Christian Theology of the Land.* Nashville: Abingdon Press, 1989.

Lyons, J. A. *The Cosmic Christ in Origen and Teilhard de Chardin: A Comparative Study.* London: Oxford University Press, 1982.

Maloney, George A., S.J. *The Cosmic Christ: From Paul to Teilhard.* New York: Sheed and Ward, 1968.

McDaniel, Jay B. *Earth, Sky, Gods, and Mortals: Developing an Ecological Spirituality.* Mystic, CT: Twenty-Third Publications, 1990.

———. *Of God and Pelicans: A Theology of Reverence for Life.* Louisville: Westminster/John Knox Press, 1989.

McDonagh, Sean. *The Greening of the Church.* Maryknoll, NY: Orbis Books, 1990.

———. *To Care for the Earth: A Call to a New Theology.* Santa Fe: Bear and Company, 1986.

McMullin, Ernan, ed. *Evolution and Creation.* Notre Dame: University of Notre Dame Press, 1985.

Merton, Thomas. *Gandhi on Non-Violence.* New York: New Directions, 1964.

Moltmann, Jürgen. *God in Creation.* San Francisco: HarperCollins, 1985.

Mooney, Christopher F. *Teilhard de Chardin and the Mystery of Christ.* New York: Harper & Row, 1964.

Muir, John. (Peter Browning, ed.) *John Muir in His Own Words.* Lafayette, CA: Great West Books, 1988.

Murphy, Roland E. *Wisdom Literature and the Psalms.* Nashville: Abingdon Press, 1983.

O'Connor, Kathleen M. *The Wisdom Literature.* Wilmington: Michael Glazier, 1988.

Perkins, Pheme. *The Book of Revelation.* Collegeville Bible Commentary. Collegeville: The Liturgical Press, 1991.

Plant, Christopher, and Judith Plant. *Turtle Talk: Voices for a Sustainable Future.* Santa Fe: Bear and Company, 1990.

Plastarus, James. *The God of Exodus.* Milwaukee: Bruce, 1966.

Presbyterian Eco-Justice Task Force. *Keeping and Healing the Creation.* Louisville: Committee on Social Witness Policy, Presbyterian Church, 1989.

Rae, Eleanor, and Bernice Marie-Daly. *Created in Her Image.* New York: Crossroad, 1990.

Schmitz-Moormann, Karl. "Theology in an Evolutionary Mode," *Zygon: Journal of Science and Religion,* Vol. 27, No. 2, June 1992.

Seed, John, et al. *Thinking Like a Mountain*. Philadelphia: New Society Publishers, 1988.

Senior, Donald. *Jesus*. Mahwah, NJ: Paulist Press, 1992.

Sheldrake, Rupert. *The Rebirth of Nature: The Greening of Science and God*. New York: Bantam Books, 1991.

Steffen, Lloyd H. "In Defense of Dominion," *Environmental Ethics*, Vol. 14, No. 2, Spring 1992.

Swimme, Brian, and Berry Thomas. *The Universe Story*. San Francisco: HarperSanFrancisco, 1992.

Teilhard de Chardin, Pierre. *Hymn of the Universe*. New York: Harper & Row, 1961.

Theissen, Gerd. *Biblical Faith: An Evolutionary Approach*. Philadelphia: Fortress Press, 1985.

Thoreau, Henry David. *Excursions*. Boston: Ticknor and Fields, 1863.

Von Rad, Gerhard. *Old Testament Theology*. Vols. I & II. Nashville: Abingdon Press, 1962, 1965.

Von Rad, Gerhard. *Wisdom in Israel*. Nashville: Abingdon Press, 1972.

Wilkenson, Loren, ed. *Earthkeeping in the Nineties: Stewardship of Creation*. Grand Rapids: William B. Eerdmans, 1991.

INDEX OF SCRIPTURAL REFERENCES

OLD TESTAMENT

NEW TESTAMENT

GENERAL INDEX